I0052670

RESILIENCE BY DESIGN

STRATEGIC RISK-PROFILING TECHNIQUES TO REFINE YOUR ORGANISATION'S CYBERSECURITY AND RESILIENCE STRATEGY

YVONNE SEARS

CONTENTS

INTRODUCTION

About the author 3
Acknowledgements 5
Preface 7

CHAPTER OVERVIEW

Chapter overview 15

1. Understanding risk profiling 19
2. Importance of risk profiling in cybersecurity 29
3. Understanding the organisation 35
4. Use of technology 45
5. Human factors 51
6. External dependencies 63
7. Industry benchmarking 69
8. Understanding stakeholder needs and expectations 73
9. Adoption of artificial intelligence 79
10. AI governance 85
11. Regulatory compliance 95
12. Strategic cyber threat intelligence 101
13. Risk appetite and tolerance 125
14. Using risk profiling to enhance incident response 131
15. The board's role 135
16. Continual improvement 141
17. Risk profiling for SMEs 147

Conclusion   153
Appendix A: 155
Appendix B: 165
Appendix C: 169
Appendix D: 173
Appendix E: 177
Appendix F: 181
Appendix G: 185
Appendix H: 189
Appendix I: 191
Appendix J: 195
Appendix K: 197
Appendix L: 199
Appendix M: 201
Appendix N: 203
Appendix O: 205
Appendix P: 207
Appendix Q: 211
Appendix R: 213
Appendix S: 215
Appendix T: 217
Appendix U: 219
Appendix V: 221
Appendix W: 225

INTRODUCTION

ABOUT THE AUTHOR

Yvonne Sears is a strategic adviser, author, and practitioner with over 25 years of experience in cybersecurity, business continuity, and privacy. She is the founder of Elev8 Resilience, a firm dedicated to helping organisations across sectors build practical, sustainable, and fit-for-purpose resilience programs. Yvonne has worked with government agencies, critical infrastructure providers, and private enterprises across Australia and internationally, delivering frameworks and strategies that translate cyber and operational risk into meaningful business value.

Throughout her career, Yvonne has been known for her ability to bring clarity to complexity, and for helping boards, executives, and operational teams navigate evolving threats, competing priorities, and regulatory expectations. She is a passionate advocate for making cybersecurity approachable and for embedding it as a natural part of how a business runs, not just as a technical add-on.

"After more than two decades in the industry, I wrote this book to share what I've learned, not just through frameworks, but through real conversations, projects, and challenges." Yvonne says. "My hope is that it sparks ideas, prompts dialogue, and helps others see how well business continuity, privacy, and cybersecurity can work together when they're aligned to what the business actually needs."

This book reflects Yvonne's core belief: cybersecurity must make sense to the business, its people, its goals, and its realities. Whether an organisation is highly regulated or just starting its cyber journey, risk profiling is the essential first step to building that understanding. It's not about ticking boxes - it's about making informed, confident decisions, every day.

Yvonne continues to mentor leaders, contribute to reform initiatives, and support organisations across Australia, the Middle East, and Asia as they build capability and culture in cyber resilience.

ACKNOWLEDGEMENTS

This book has been years in the making, shaped by my decade-long drive to capture the lessons, patterns, and practical wisdom I've gathered as a career consultant. Working across industries, organisations, and complex business challenges has given me a unique lens on risk—one that I've tested, challenged, and refined over time. I've seen what works, what fails, and where many still struggle to turn risk management into something meaningful.

This book wouldn't exist without the people who have supported me not just in this project, but across my entire professional journey.

First and foremost, to my husband, Pete: your unwavering belief in me has been a constant source of strength. You made sacrifices most wouldn't, giving up your own career so I could pursue mine. You've stood by me through the long hours, the travel, the career pivots, and now this book. I'm endlessly grateful.

To the Elev8 Resilience community and my professional peers: thank you for the encouragement, conversation, and collaboration that have fuelled much of what's in these pages. Elev8 has been more than a business—it's been a platform to challenge thinking, push boundaries, and reframe how we approach cybersecurity, risk, and resilience.

And to those who generously gave their time to review early chapters and provide honest, thoughtful feedback: your insights have shaped the clarity and strength of this work.

PREFACE

Every business, no matter what industry—whether mining, transportation or construction management—needs to appreciate how information security risk management plays a crucial role in daily operations. It's not just for heavily regulated organisations such as healthcare or finance. Today every business relies on data. When that data is compromised—whether through loss, manipulation, or unauthorised access—the impact can be severe.

Data drives business intelligence, efficiency, and performance. It's the lifeblood of any industry and protecting it should be non-negotiable. Imagine if a mining company relied on faulty data and drilled in the wrong location, even by a small margin. The financial loss from wasted time, resources, and missed revenue opportunities could be in the millions. Or think of a construction project brought to a halt because the scheduling system went down. The consequences would be costly and far-reaching.

I've spent years observing how businesses approach information security and cybersecurity, terms often used interchangeably but referring to a broad spectrum of digital and data protection. For the sake of simplicity, I'll refer to both under the umbrella of "cybersecurity" throughout this book. Too often, I've seen it treated as an afterthought or written off as a cost centre. But when done well, cybersecurity is far more than a defence mechanism—it can be a powerful business enabler, helping companies gain a competitive edge, unlock new opportunities, build customer trust, and position them for long-term success. That's why I founded Elev8 Resilience and why I'm writing this book. My goal is to empower all businesses to grasp how cybersecurity fits into their operations, moving it from a mere obligation into a core part of their strategy, culture, and value proposition.

WHY SHOULD YOU READ THIS BOOK?

If you're tired of cybersecurity being seen as a technical burden rather than a business enabler, this book will fundamentally change how your organisation approaches risk management. Whether you're struggling to translate vulnerabilities into business language or justify security investments to leadership, or simply don't know where you're most exposed, you'll discover a practical framework that transforms uncertainty into confidence. You will learn to identify genuine business risks, articulate the real cost of inaction, and position security as a strategic advantage that drives growth and competitive differentiation.

MY GOAL

I want to help 200 Australian mid-market and specialist small and medium enterprise (SME) leaders transform their approach to cyber risk management, enabling them to achieve measurable return on investment (ROI) improvements and confident risk prioritisation within 12 months of reading this book.

YOUR RISK-PROFILING JOURNEY CONTINUES ONLINE

This book is designed to be your practical companion for embracing and implementing risk profiling within your organisation. However, your journey doesn't end on the final page. I've created a comprehensive suite of digital resources specifically for readers who are ready to transform theory into action.

These resources are designed to accelerate your implementation, provide immediate value to your stakeholders, and confirm you can confidently navigate the complexities of modern risk management. Whether you're a seasoned Chief Information Security Officer (CISO) seeking additional tools or a business leader taking your first steps into structured risk management, these companion resources will support your success.

Ready to elevate your risk-profiling approach?

Access your reader resources at **www.elev8ciso.com** and join a community of practitioners who are transforming the ways their organisations acknowledge and manage risk.

RISK PROFILING

I've chosen to dedicate my first book to risk profiling because it's the foundation upon which every business must build its understanding of vulnerabilities. Risk profiling isn't just a preliminary step—it's an essential process that shapes how a company identifies and prioritises its greatest risks. By focusing on risk profiling, businesses can pinpoint where they are most exposed, enabling them to align their resources and strategies accordingly.

In the ever-evolving world of cyber threats and digital transformation, risk profiling is at the forefront of cyber threat intelligence (CTI) and strategic planning. It goes beyond merely identifying threats—it provides critical insight into the specific activities, dependencies, and complexities unique to each business. While this book focuses its scope on the context of cybersecurity, the principles discussed can help organisations of all sizes and industries create a resilient and proactive approach to managing other key operational risks.

The current threat reality

Australian businesses are confronting an unprecedented convergence of threats that traditional risk frameworks struggle to address comprehensively:

- Ransomware with double extortion has evolved beyond simple encryption, with threat actors now combining data theft, supply chain targeting, and regulatory pressure tactics.
- Artificial intelligence (AI)-powered attacks are enabling threat actors to conduct more convincing social engineering campaigns and automate vulnerability discovery at scale.
- Supply chain compromises are creating cascading risks that traditional asset-based risk assessment fails to capture adequately.
- Nation-state threats are increasingly targeting critical infrastructure and defence contractors, blurring the lines between cybercrime and geopolitical risk.

The limitations of traditional risk profiling

Conventional risk assessment approaches typically focus narrowly on:

- Static asset inventories that fail to account for dynamic cloud environments
- Isolated threat categories that don't reflect the interconnected nature of modern attacks
- Historical incident data that provides limited insight into emerging attack vectors
- Compliance-driven frameworks that prioritise regulatory requirements over business resilience.

These approaches create dangerous blind spots in an organisation's risk awareness, particularly around:

- Emerging technology risks: AI implementation, Internet of Things (IoT) proliferation, and cloud migration create new attack surfaces that traditional frameworks don't adequately address.
- Third-party ecosystem vulnerabilities: modern business operations depend on complex supplier networks that extend far beyond traditional vendor risk assessment.

- Adaptive threat actor behaviour: sophisticated adversaries continuously evolve their tactics, techniques, and procedures (TTPs) faster than static risk assessment can capture.

The business impact of inadequate risk profiling

The consequences of relying on outdated risk-profiling approaches are increasingly severe:

- Misallocated security investments: resources are directed towards lower priority risks, while critical vulnerabilities remain unaddressed.
- Regulatory compliance failures: traditional approaches may satisfy basic compliance requirements while overlooking emerging regulatory expectations around AI governance and supply chain security.
- Business disruption: inadequate comprehension of interconnected risks leads to cascading failures during incidents.
- Competitive disadvantage: organisations with superior risk knowledge can pursue digital transformation initiatives more confidently and effectively.

The integrated risk-profiling imperative

Today's threat environment demands a fundamentally different approach to risk profiling—one that integrates cybersecurity, business continuity, privacy, and emerging technology governance into a unified risk perspective. It is essential to incorporate real-time threat intelligence, ensuring risk assessments reflect current adversary capabilities and intentions. Rather than focusing solely on technical controls, this approach considers business resilience outcomes and adapts dynamically to evolving threat landscapes and business environments.

This book presents a comprehensive methodology for modern risk profiling that directly addresses these challenges. Instead of treating cybersecurity, business continuity, and privacy as isolated disciplines,

11

the method described here recognises that effective risk management in 2025 requires integrated thinking, mirroring the interconnected nature of today's business operations and threat vectors.

The frameworks and strategies outlined in the following chapters are based on extensive experience with Australian organisations across government, financial services, healthcare, and technology sectors. They reflect not only the latest best practices but also anticipate the risk-management challenges organisations will encounter as digital transformation continues to accelerate.

BOOK STRUCTURE

This book is designed to be a practical companion, a tool you can use, revisit, and rely on whether you're starting from scratch or strengthening your existing risk-management approach. It's structured to help you take real action—whether that means initiating a risk-profiling process, refining your strategy, or supporting others in your organisation to better comprehend the risks they face.

Each chapter builds upon the previous ones, offering a natural progression for those reading cover to cover. But equally, you can jump straight to the section most relevant to your current needs, whether you're deep in a project, preparing for an audit, or making a strategic decision. Risk management isn't always linear and this book reflects that reality.

Throughout, you'll find real-world examples, practical templates, and reference tools designed to save you time and help you make confident decisions. These resources are also consolidated in the final chapter so you can quickly locate and reuse them as needed.

Whether you keep this book on your desk or in your bag, or bookmarked digitally, it's meant to be used—not just read. My goal is for this to be a trusted resource you return to at every stage of your journey towards stronger, smarter risk profiling.

CHAPTER OVERVIEW

This book is designed to guide you through the critical process of risk profiling, specifically within the realm of cybersecurity. Each chapter will take you deeper into the components and strategies that shape cyber risk management, helping your business better comprehend its vulnerabilities and how to address them. Here's what you can expect:

Chapter 1: Understanding risk profiling

We begin by exploring what risk profiling is and why it's the first essential step in any comprehensive cybersecurity strategy. This chapter introduces the core concepts and frameworks needed to appreciate how vulnerabilities can be mapped and prioritised.

Chapter 2: Importance of risk profiling for cybersecurity

This chapter delves into how risk profiling fits into the broader field of CTI. We'll examine how defining your risk profile allows you to foresee potential cyber threats and prepare strategic defences.

Chapter 3: Understanding the organisation

Every business is unique and risk profiles should reflect that. In this chapter, we discuss how to customise your risk profile to your company's specific operational structure, dependencies, and threat landscape.

Chapter 4: Use of technology

This chapter explores 7 distinct IT environment categories—ranging from highly agile and innovative setups to legacy and shadow IT—each presenting unique challenges and implications for risk. Through examining the structure, complexity, and governance of IT systems, you are guided to uncover hidden vulnerabilities, assess technical debt, and identify critical operational dependencies.

Chapter 5: Human factors

Every business is unique in how staff, customers, and suppliers engage with its products and services. In this chapter we look deeper into profiling risks related to insider threats.

Chapter 6: External dependencies

Third-party relationships are foundational to modern business operations, but they also introduce significant risk that must be understood and managed proactively. This chapter outlines how to evaluate and profile external dependencies by building a comprehensive third-party register, drawing on cross-functional input from finance, procurement, IT, and operational risk teams.

Chapter 7: Industry benchmarking

Every industry faces unique risks based on the nature of its services, regulatory landscape, customer expectations, and reliance on technology. In this chapter, we explore how different sectors experience and manage cyber threats, covering finance, healthcare, autonomous & electric vehicle (EV), mining & natural resources, and education & research industries. By defining the nuances and threat profiles of each, you'll gain insight into how industry context shapes your risk profile and how to benchmark your approach against peers.

Chapter 8: Understanding stakeholder needs and expectations

A successful risk-profiling process begins with examining who your stakeholders are and what matters most to them. Whether customers, regulators, executives, or operational teams, their needs and expectations shape your organisation's risk priorities. This chapter guides you through identifying your key stakeholders and uncovering what they value, laying the foundation for a resilient and aligned cybersecurity strategy.

Chapter 9: Artificial intelligence

Emerging technologies bring new challenges and risks. We examine how AI systems are shaping organisation risk profiles and what businesses can do to prepare for these next-generation threats.

Chapter 10: AI governance

Effective AI governance establishes ethical, safe deployment that aligns with organisational risk tolerance and regulations. This chapter shows how AI governance requires cross-functional collaboration and new roles to promote accountability, explainability, and resilience throughout the AI system lifecycle.

Chapter 11: Regulatory compliance

Regulations are formed to give direction on what is acceptable behaviour and to set community expectations in the management of risks. It would be remiss of me not to look in to how regulations play their part in the risk profile of your organisation.

Chapter 12: Strategic cyber threat intelligence

In today's interconnected world, the risks posed by third-party vendors and supply chains are critical. This chapter focuses on how to extend your risk-profiling process to include external partners and examine the vulnerabilities they introduce into your environment.

Chapter 13: Risk appetite and tolerance

Risk profiling is incomplete without clarity on how much risk your organisation is willing and able to accept. In this chapter, we explore the concepts of risk appetite and risk tolerance—two often misunderstood but essential components of strategic decision-making. You'll learn how to define these boundaries in practical terms and how to use them to guide investment, response planning, and stakeholder engagement.

Chapter 14: Using risk profiling to enhance incident response

Resiliency culminates in the ability to respond to and recover from a risk event, so in this chapter we consider how an effective risk profile can inform cyber incident response planning.

Chapter 15: The board's role

Risk profiling is more than just an "IT issue" —it's a business enabler. We'll conclude by examining the way aligning your risk profile with your overall business strategy can create a stronger, more resilient organisation.

Chapter 16: Continual improvement

Risk profiles are not static—they need to evolve as your business grows and as new threats emerge. Here, we look at how to continuously update and adapt your risk profile to stay ahead of the curve.

Chapter 17: Risk profiling for SMEs

In this chapter we explore some quick wins and strategies for SMEs in how they too can determine and manage their risk profiles with limited resources.

Conclusion

As the final pages unfold, the book calls readers to action, urging organisations to embrace risk profiling not as a one-off exercise, but as a continuous commitment woven into the fabric of their strategy and culture. Leaders and teams alike are challenged to view uncertainty as a catalyst for innovation and growth, leveraging the principles and practices outlined within to build resilience and sharpen their competitive edge.

Rather than waiting for risk to materialise, the message is clear: proactively cultivate the capability to anticipate, adapt, and thrive—making risk profiling an integral driver of organisational success in an unpredictable world.

Appendices

The appendices serve as a practical companion to the main text, offering a curated collection of templates and real-world examples. Readers will find step-by-step guides and tried-and-tested frameworks designed to help them apply risk profiling concepts directly to their own organisations. Each template is structured for clarity and ease of use, enabling leaders and teams to work through the exercises at their own pace. Whether refining a risk profile, mapping out governance strategies, or stress-testing business continuity plans, the appendices provide actionable resources for embedding resilience and strategic foresight into everyday practice.

"Giving up the illusion that you can predict the future is a very liberating moment. All you can do is give yourself the capacity to respond to the only certainty in life—which is uncertainty. The creation of that capability is the purpose of strategy."

Lord John Browne, former Group Chief Exec of BP

CHAPTER 1

UNDERSTANDING RISK PROFILING

Governance, risk, and compliance (GRC) in cybersecurity are relevant across various business sectors and can be applied to organisations of any size. When implemented effectively, they shape strategies that align with an organisation's mission and objectives.

A practical approach begins by assessing the specific challenges facing the organisation, rather than focusing solely on predetermined goals. Identifying and defining core issues allow for the development of a strategic plan aimed at addressing the most significant challenges.

Risk profiling plays a central role in this process. Drawing from experience across industries and organisations of different sizes, risk profiling consistently serves as an effective starting point, despite varying requirements for security controls and budgets.

WHAT IS RISK PROFILING?

Risk profiling sets the stage for strategic planning, enabling a deep comprehension of an organisation's needs and expectations. It provides insight into how and where the organisation operates, who its stakeholders are, and the potential impacts of disruptions to its products and services. More than a cornerstone of cyber resilience, risk profiling is a holistic tool that informs strategic decisions across financial, operational, and regulatory landscapes. When done effectively, it can uncover opportunities for process improvement, identify efficiencies, and enhance competitive advantage.

Key components of risk profiling include internal and external factors, technology usage, and cybersecurity risks. While we delve deeper into these through the course of the book, it's important to note here that developing a robust cybersecurity strategy also requires comprehensively understanding how the business operates—its financial exposures, current or planned changes (such as organisational or digital transformation), and geopolitical or geographical risks that could introduce new vulnerabilities.

At its core, cybersecurity is a strategic pillar for preventing loss and ensuring the continuity of business processes, products, and services. Business continuity and resilience management, therefore, sit naturally at the top of the GRC structure. Without a clear grasp of why loss prevention is critical, securing commitment from executive teams and stakeholders for cybersecurity investments becomes significantly more challenging.

Risk profiling overcomes this challenge by creating a common language for discussing risks, bridging gaps between technical, operational, and executive teams. This alignment fosters greater awareness and cooperation across the organisation. By speaking the language of the business, risk profiling highlights dependencies, vulnerabilities, and potential threats to operations, ensuring stakeholders recognise its value.

Moreover, risk profiling is inherently collaborative. It requires input from every layer of the organisation to build a comprehensive and actionable profile of vulnerabilities and opportunities. This collective effort not only strengthens the risk profile's accuracy, but also encourages buy-in by making all stakeholders active participants in the process.

In summary, risk profiling provides a structured process for identifying and prioritising risks that could impact the organisation. It is not a one-time activity, but a dynamic and ongoing process. In an era of rapid technological advancement and global interconnectivity, risks are constantly evolving—making regular reviews essential to maintaining relevance and resilience.

As we explore the tools, techniques, and real-world applications of risk profiling in subsequent chapters, this foundation will guide you in developing a resilient and adaptive strategy for your organisation.

WHY DOES IT MATTER?

Risk profiling is the groundwork for prioritising efforts and resources, ensuring they are allocated to address the risks that pose the greatest threats to the organisation's objectives. It shifts the focus from reactive to proactive risk management, enabling security managers and risk practitioners to craft targeted mitigation strategies that not only reduce exposure, but also create measurable value through ROI.

By defining the organisation's risk appetite and tolerance levels, risk profiling empowers decision-makers to navigate the trade-offs between opportunity and security. It allows leaders to make calculated decisions, knowing which risks can be accepted and which must be either mitigated or avoided altogether.

Moreover, risk profiling fosters accountability and transparency. It provides a clear framework for justifying investment in security and resilience initiatives, making it easier to secure executive and stake-holder buy-in. This clarity strengthens governance and demonstrates that risk management is integral to achieving the organisation's strategic goals.

Finally, in a landscape of constant change—whether driven by technological advancements, shifting market conditions, or regulatory demands—risk profiling equips organisations with the agility to adapt. It ensures that as the organisation evolves, so too does its comprehension of its threats and vulnerabilities, keeping the business resilient and future ready.

Issues

I've put together my thoughts and experiences in this book primarily to address the following issues I've seen time and time again over the past 25 years! Risk profiling is essential to gain clarity throughout your organisation and my hope is that, by investing time in this, you can manage the following issues and constraints seen so often.

Issue	Result
The organisation has no clarity or comprehension of its risk profile in terms of cybersecurity.	There is a lack of support or funding for the cybersecurity strategy, and/or lack of oversight of key risk indicators (KRIs) to enhance the organisation's ability to anticipate and therefore prevent events occurring.
There is no clarity on what data is processed, why, and where it is stored.	You can't protect what you can't see! If there is no clarity over the type and value of data processed, then it's not possible to define the required controls needed to protect it.
There is a lack of awareness or governance around legacy systems and "technical debt" and therefore a lack of understanding of operational risks.	Without appreciating the IT dependencies and complexity of the technical environment, it's not possible to design a fully secure environment or transition away from legacy systems which limit the latest cybersecurity controls.
There is over-reliance on third parties and an assumption that they	This leaves the organisation exposed to cybersecurity risk events that occur via a third party,

Issue	Result
"have security covered" with little to no validation or assurances.	whether user non-compliance or a hacker "hopping" from one organisation to another.
It is assumed IT "owns" all cybersecurity risks.	This limits the conversation around information security and cybersecurity to "just IT". Many risk events are not "just" caused by technical failures or vulnerabilities, but stem from how business processes are designed and the culture across the organisation. Making information security and cybersecurity a purely IT issue exposes the organisation to unnecessary risks.

HOW IS RISK PROFILING DISTINCT IN CYBERSECURITY VERSUS GENERAL RISK MANAGEMENT?

When evaluating a risk profile with cybersecurity in mind, certain nuances set it apart from the standard operational risk-management approach. While traditional risk management focuses broadly on operational, financial, and strategic risks, cybersecurity risk profiling zooms in on digital ecosystems, threat landscapes, and technical resilience. It demands an iterative, data-driven, and collaborative approach to guarantee the organisation stays ahead of evolving threats.

Here are 8 key distinctions and considerations for you to consider:

1 Nature of cybersecurity risks

Unlike traditional risks, cyber risks are highly dynamic and have greater velocity, with new threats emerging constantly (e.g., ransomware, zero-day vulnerabilities). Risk profiling in cybersecurity requires continuous monitoring and updates to remain relevant.

Many cybersecurity risks are the result of external threat actors, such as hackers, nation-states or state sponsors, or based on the outcome of behaviour—making them more unpredictable or intentional than natural or operational risks.

Additionally, the digital nature of cybersecurity means threats can originate from anywhere in the world, adding a layer of complexity to identifying and mitigating these risks.

2 Focus on data and technology

Cyber risk profiling focuses heavily on acknowledging the organisation's digital assets (e.g., data, systems, networks) and their criticality. Questions like "What are our crown jewels?" and "What systems are indispensable?" are central to correctly scoping and prioritising resources.

Cybersecurity risks often extend beyond the organisation to third-party vendors, cloud providers, and supply chains, requiring a broader scope of profiling.

Risk profiling in cybersecurity must account for vulnerabilities in software, hardware, and configurations, which are not typically part of general risk assessment.

3 Probability and impact measurement

Cyber risks often hinge on whether a vulnerability can and will be exploited by a threat actor. We need to consider both the motives and the capabilities of the threat actor in evaluating likelihood. This adds an additional layer of analysis compared to risks that may arise naturally.

Cyber incidents can have cascading effects across operations, finances, reputation, and compliance, with the impacts being felt across multiple stakeholders (partners, supply chains, and clients). Risk profiling must consider not only direct losses, but also secondary and tertiary impacts (e.g., regulatory fines, loss of customer trust).

· · ·

4 Regulatory and legal obligations

Cyber risk profiling must align with frameworks such as ISO/IEC 27001 Information Security, Cybersecurity and Privacy Standard, APRA Standards (CPS 230/234), the National Institute of Standards and Technology (NIST), or local laws such as Australia's Cyber Bill and Europe's General Data Protection Regulations (GDPR). These add unique parameters to the risk-evaluation process.

Profiling also needs to account for risks like data breaches that could lead to lawsuits, penalties, or breaches of contract.

5 Risk appetite in cybersecurity

Organisations may have very low appetites for risks related to customer data, intellectual property (IP), or critical infrastructure, distinguishing cyber risk appetite from broader risk categories.

Equally, cyber risk profiling must ensure security measures do not stifle innovation or business agility and this is where defining the organisation's appetite, or tolerance, for risk plays an important role in risk-treatment decisions.

6 Stakeholder perspectives and collaboration

Cyber risks require collaboration across technical teams, legal departments, compliance, and the board. Each stakeholder has different risk priorities that must be reconciled. It's essential to understand the needs and expectations of all stakeholders so that their perspectives are considered when evaluating and communicating cybersecurity risks.

Unlike general risks, cybersecurity often involves technical jargon. Effective profiling requires translating this into business terms that resonate with leadership and non-technical stakeholders.

7 Scenario planning and incident response

Cyber risk profiling emphasises the most relevant potential attack scenarios (e.g., phishing campaigns, insider threats) and their specific impacts on the organisation.

Cyber risk profiling also considers an organisation's readiness for incidents, ensuring the organisation has robust detection, response, and recovery mechanisms in place.

8 Cybersecurity-specific metrics

Metrics like mean time to detect (known as MTTD) and mean time to recover (known as MTTR) are critical in cyber risk profiling and distinct from general risk-management KPIs. Metrics may also consider the velocity or scale of an attack to measure consequence.

Cyber risk assessments often use specialised tools to score risks based on threat intelligence, vulnerability data, and asset criticality.

"Strategic risk is a matter of choice.
Operational risk is a matter of execution."

Douglas W. Hubbard, leading consultant, speaker, and author

CHAPTER 2

IMPORTANCE OF RISK PROFILING IN CYBERSECURITY

Many organisations still try to squeeze cybersecurity risks into a corporate risk framework often designed with operational, financial, or compliance risks in mind. This approach fails to capture the complexity, dynamism, and technical nuance of cybersecurity threats.

Although progress has been made in improving consequence definitions to better reflect cyber-related impacts, cybersecurity teams often encounter challenges in trying to harmonise risk ratings with models that were not specifically designed for them. This can result in poor or no alignment between cybersecurity teams and decision-makers.

When there is no direct connection between cybersecurity risks and organisational goals, discussions may lose momentum. Cyber risks are sometimes viewed as hypothetical or low priority, particularly when assessed mainly from financial or regulatory perspectives.

For example, one mining organisation's enterprise risk framework was primarily focused on work health and safety, environmental compliance, and production continuity, areas where its risk-management approach had matured over time. Within this framework, it was difficult for any cybersecurity risk to be classified higher than medium. As a result, despite the changing threat landscape, executives frequently gave lower priority to cyber risks, indicating that if an incident occurred, recovery costs would simply be covered. The primary focus remained on direct financial considerations and safety metrics.

The root cause wasn't just a lack of cybersecurity awareness. It stemmed from:

- No dedicated cybersecurity risk-appetite statement
- Oversimplified likelihood definitions offering only high-level categories like "unlikely," "possible," and "highly likely" with arbitrary probabilities attached
- No mechanism to represent or justify technical or threat intelligence inputs.

WHY CYBERSECURITY REQUIRES A DIFFERENT LENS

Cybersecurity risks are multidimensional. Unlike conventional risks, likelihood assessments must draw from several interrelated factors, including:

- Control confidence: informed by control validation, testing, and maturity assessment
- Threat intelligence: including ease of exploitability, capability of threat actors, and known campaigns
- Incident history: near misses, past breaches, and emerging threat trends
- Frequency: ranging from daily threats (e.g., phishing) to rare but catastrophic events.

Most importantly, we must consider velocity: the speed at which a cyber event can escalate once triggered. Cyber incidents can unfold in just minutes or hours, quickly outpacing response capability and causing disproportionate damage before mitigation is even possible.

A ransomware attack that spreads within an hour, exfiltrates data, and cripples operations illustrates this risk velocity. Traditional frameworks often ignore this dimension, leaving executives blind to the time-critical nature of cyber risk.

Until organisations define a clear risk profile for cybersecurity, one that aligns with their operational context, risk managers will continue to struggle to articulate threats in terms leadership can act on. This is exactly what ISO/IEC 27001:2022 Clause 4, Context of the Organisation, seeks to address. It requires organisations to:

- Identify internal and external issues that affect cybersecurity
- Consider the needs and expectations of stakeholders
- Align the information security management system (ISMS) with broader business objectives and risks.

This contextual foundation is essential. It upholds the requirement that cybersecurity is embedded not just in policy but in purpose—treated as an enabler of organisational resilience, not a bolt-on function.

ISO 27001: CONTEXT OF THE ORGANISATION

It is no accident that setting context for cybersecurity is at the forefront of the standard's requirements. This means providing the context in which the required security controls and capabilities are to be implemented and identifying the security principles the client aspires to achieve.

Its primary purpose is to justify the priorities within the cybersecurity program and it helps to set the scope of work. The context should be clear, precise, and relevant to the client.

A context document is not only a mandatory requirement for ISO 27001 certification, but can often be used to support a business case for change (e.g., across business continuity, cybersecurity, and privacy management) as it defines:

- Organisation overview: what does the organisation do, why, and how?
- Stakeholders: who depends on the products and services, what is the impact of loss, who has a vested interest in the success of the ISMS and what are their needs and expectations?

- Dependencies: what does the ISMS depend upon to succeed and what if these factors are not available?
- Overview of standards, regulations and legislation: in terms of information security, cybersecurity and privacy, what regulations impact the ISMS, what are the specific requirements, and how will they be met by the ISMS?
- Risk profile: essentially, why is cybersecurity important to the organisation?
- PESTEL analysis: internal and external impacts to the ISMS
- Summary: high-priority/focus areas to lead the strategic plan.

"There are only two types of companies:
those that have been hacked, and those that will be."

Robert Mueller, former Director of the FBI

CHAPTER 3
UNDERSTANDING THE ORGANISATION

A common starting point for designing a security strategy is to learn about the organisation's structure, culture, and work practices. Without this defined, it is difficult to create a cybersecurity strategy that meets the organisation's needs and expectations.

BACKGROUND, VISION, AND OBJECTIVES

Investing time in understanding the business is foundational. Uncover its "secret sauce"—the unique factors that differentiate its offerings—and analyse how these are created and delivered to customers.

- Why does the business exist?
- What problem does it solve for its customers or stakeholders?
- How do its core processes and values align with its mission?

Start by studying the strategic plan, mission, and vision statements. These documents offer insight into the organisation's goals and language, helping tailor a cybersecurity strategy that aligns with and enhances its core objectives.

For example:

- If growth through digital transformation is a key objective, how can cybersecurity mitigate risk while enabling innovation?
- How can a robust cyber strategy unlock new opportunities such as customer trust, regulatory compliance, or operational efficiency?

In terms of the organisation's background, investigate its historical milestones and its culture. Has it faced challenges like breaches, operational disruptions, or regulatory issues? How did it respond? What was the outcome? Is it risk averse, or innovation driven?

DOES IT PROVIDE SERVICES, CREATE PRODUCTS, OR BOTH?

Clearly defining the organisation's offerings and what makes it unique is essential for mapping dependencies and vulnerabilities. Assess the full value chain by asking:

- What tangible products or intangible services are delivered?
- What are the end-to-end processes that support these offerings?
- Who are the key stakeholders (customers, partners, regulators, employees) and how do they interact with the organisation?

If the organisation has adopted quality-management standards like ISO 9001, this information should be readily available. If not, start mapping out the structure:

- What are the critical business processes?
- What triggers internal processes?
- What happens within the organisation to act on those triggers?
- At what point is the product or service created and what are the critical stages in the process?

Analyse the digital components of these processes: are cloud systems, AI tools, or third-party vendors involved?

This process reveals potential "pain points" or vulnerabilities, especially where dependencies, bottlenecks, or external touchpoints exist. Verify your findings with internal stakeholders to refine the map, identifying:

- Criticality: which steps are essential for business continuity?
- Risks: where are the vulnerabilities that overlap with cyber concerns?

INDUSTRY AND STAKEHOLDER IMPACTS

What role does the organisation play in its industry and supply chain?

The organisation's position in the supply chain influences its risk profile. Questions to ask include:

- Is the organisation a critical provider of goods or services within its industry?
- What impacts would delays, disruptions, or loss of service have on customers, regulators, and partners?
- Are there reputational risks tied to the organisation's role or industry?

Mapping out these dependencies helps refine the organisation's risk appetite by quantifying impacts across several dimensions:

- Stakeholders: customers, investors, regulators, and partners
- Reputation: public perception and brand trust
- Continuity: operational ability to deliver during disruptions
- Compliance: adherence to industry-specific regulations

UNDERSTANDING ORGANISATIONAL ANATOMY

Now that you comprehend the makeup of the organisation at a high level and its place in the world, explore the functions of the organisation that bring it to life:

Departmental structure	Identify all departments and supporting business units that make up the organisation. Map their relationships and dependencies to discern how they interact.
Geographical considerations	Document where functions are located and analyse any environmental, geopolitical, or economic factors that could influence the risk profiles of these locations.
Specialist resources	Determine dependencies on specialised skill sets or equipment that are critical to product and service delivery. These represent potential single points of failure.
Unique attributes	Identify any unique organisational characteristics that may impact resilience or create distinctive risk factors requiring special attention.

Business processes

After identifying organisational functions, the next critical step is mapping the underlying business processes in detail. The supplier, input, process, output, client (SIPOC) methodology provides a structured framework for visualising process flows and their dependencies. This comprehensive process mapping serves as the foundation for identifying operational risks and their potential impacts:

Suppliers	These are the entities providing inputs needed for the process to function, including internal departments, external vendors, or service providers.
Inputs	These are materials, information, authorisations, or resources required before the process can begin or continue.
Process steps	These are the sequence of activities that transform inputs into outputs, focusing on critical path actions.
Outputs	These are the products, services, information, or decisions created through the process.
Clients	These are the recipients who depend on the process outputs, whether internal stakeholders or external customers.

When creating SIPOC diagrams for each business process, maintain a balance between comprehensive detail and useable clarity. Begin with major processes before drilling down into sub-processes. Involve process owners and frontline staff to maintain accuracy. Use consistent terminology across diagrams to facilitate organisational understanding. Update diagrams regularly as processes evolve to maintain relevance for risk-assessment purposes.

For complex organisations, prioritise mapping of critical processes first —those directly tied to value delivery, regulatory compliance, or significant revenue generation. This focused approach ensures that your risk assessment addresses the most consequential areas of operation while building momentum for comprehensive mapping.

Dependency mapping

For each process step, document all dependencies required for successful execution. This comprehensive mapping reveals potential vulnerabilities across 4 key categories:

- People dependencies include both internal staff and external contractors with specialised knowledge.
- Asset requirements encompass everything from critical manufacturing equipment to software applications and facilities.
- Knowledge dependencies include documented procedures, institutional knowledge, and specialised training.
- External dependencies cover supplier relationships, outsourced services, and regulatory requirements that influence process execution.

By mapping the processes this way it's possible to visualise the pain points of each process, recognise dependencies, and explore the operational risks associated with the activity.

Once you have established the dependency mapping, consider the impacts of loss:

Input delays	Analyse the consequences when critical inputs are delayed or disrupted. Identify upstream dependencies that could trigger cascading failures and estimate potential operational impacts.
Personnel unavailability	Assess the implications when key employees, contractors, or third-party service providers are unavailable. Identify knowledge gaps, cross-training needs, and succession-planning requirements.
Asset compromise	Evaluate scenarios where physical assets or IT systems are damaged, lost, or compromised. Determine recovery requirements, alternative solutions, and security vulnerabilities.
Document integrity issues	Consider impacts when documentation is incomplete, outdated, or compromised. Identify data-integrity risks, compliance implications, and operational disruptions from poor information management.
Supply chain disruption	Map consequences when critical services or products become unavailable. Identify single-source dependencies, alternative suppliers, and stockpiling requirements.

Use this as a discussion point with each function owner and relevant stakeholders to examine the pain points they encounter and determine whether these agree with your assessment. This also identifies any significant pain points that may require closer monitoring to prevent further issues.

The aim is to accurately map the risk profile to specific business processes so stakeholders can better understand the risks within their departments and assume responsibility for these risks.

Such discussions frequently uncover unofficial workarounds and undocumented dependencies not captured in formal documentation. Direct stakeholder involvement in the risk-identification process encourages both ownership and accountability, while providing a more complete view of operational conditions.

This approach also supports the development and maturity of the business impact analysis (BIA) process necessary for forming resilience strategies for the organisation, such as creating a new BIA or ensuring the accuracy and completeness of existing BIAs.

"Risk comes from not knowing what you're doing."

Warren Buffett, CEO of Berkshire Hathaway

CHAPTER 4
USE OF TECHNOLOGY

To comprehensively assess an organisation's risk profile, it is critical to define how it utilises technology and the nature of its IT environment. The categorisation of IT systems and practices provides valuable insights into potential vulnerabilities, technical debts, and the degree of complexity or stability within the organisation. There are 7 key areas to consider when defining the IT risk profile of the organisation:

(A) ADVANCED / CONTINUAL IMPROVEMENT / AGILE ENVIRONMENT

Organisations in this category prioritise innovation and agility, with greater tolerance for risk, but how is it controlled? They typically have dedicated technical and development teams constantly seeking the best IT solutions or creating bespoke technologies to enable the business or create new solutions for customers. While this fosters competitiveness and adaptability, the rapid adoption of new technologies or continual change can introduce:

- Increased attack surfaces from new integrations
- Higher likelihood of unpatched systems or misconfigurations
- Challenges in maintaining consistent security policies
- Greater dependencies on specialist skill sets.

If relevant to your environment, work with your IT team to investigate further how the organisation supports this way of working and how it currently mitigates potential risks. For example, has it embedded security within the development lifecycle (DevSecOps) so that there is a culture of "security-by-design", what does this look like at each stage, and how is change managed during deployment?

(B) CONSTANT TECHNICAL ENVIRONMENT WITH MINIMAL CHANGE

A stable IT environment where updates and changes are limited provides a lower risk profile. However, this environment can also foster complacency, leading to:

- Outdated security protocols
- Systems that become unsupported over time.

Discuss with key stakeholders their needs and expectations for IT services, in addition to your discussions with the IT team and Chief Information Officer (CIO), if you have one. What steps are in place to review the effectiveness and assumptions of the environment? How often do they monitor for system vulnerabilities, capacity, and lifecycle of assets? Do processes verify that updates are applied consistently and that they prevent legacy systems being created (i.e., proactively ensuring that no new systems fall outside of support)?

(C) COMPLEXITY

Complex IT environments, with multiple integrated platforms and services, often increase operational efficiency, but pose significant challenges:

- Difficulties in maintaining a clear view of the IT landscape
- Increased likelihood of vulnerabilities in integration points.

This environment is typically observed in large critical infrastructure that also supports an operational technology (OT) environment, such as mining, telecoms, and transport. It's also observed in organisations that have grown exponentially through acquisitions and mergers.

Discuss with the technical teams how they manage the overall complexity of the environment: is there a centrally managed database (CMDB), detailed system architecture diagrams, or supporting documentation? How is interoperability monitored to prevent data leaks and system failures?

(D) LEGACY SYSTEMS

Legacy systems, while supporting critical business functions, are often outdated and difficult to replace. They pose unique risks, including:

- Incompatibility with modern security tools
- High likelihood of unpatched vulnerabilities
- High likelihood of being unsupported by the vendor (i.e., no further updates or security features over time).

Although legacy systems have been accepted in this type of organisation, validate whether the executive really comprehend the associated risks in maintaining these systems longer term. What plans or strategies exist for legacy system modernisation? How has the cyber risk been mitigated to date? For example, are these systems segmented from the rest of the IT environment?

(E) SHADOW IT

"Shadow IT" is a term associated with unmanaged systems or services procured without central oversight. These environments introduce risks such as:

- Unknown vulnerabilities
- Potential non-compliance with data protection / privacy regulations.

Is the organisation even aware of the concept of shadow IT? Does it plausibly exist based on the current procurement policies and procedures, and how does the organisation identify whether it's an issue or not? If it's a possibility, how is it monitored and managed when detected?

(F) OFF THE SHELF

This is where all IT products are standard off the shelf, have not been significantly modified for the organisation, and can be easily procured again if needed. Standard, easily replaceable IT products often reduce complexity and risk. However, their reliance on vendor security can create vulnerabilities such as:

- Dependence on vendor patch cycles
- Potential exposure to widespread zero-day vulnerabilities.

Interview the IT team to understand how standard products are monitored for compliance and what controls are available to the organisation in terms of service-level agreements (SLAs) (covering system availability, security updates, and support). Have the products been evaluated for risk based on how they are used within your organisation and are the available controls appropriate for how they are used?

(G) BESPOKE

Customised or in-house-developed IT solutions provide flexibility, but can introduce unique challenges such as:

- Security risks from untested or poorly coded solutions
- Dependence on specific individuals or teams for maintenance.

Similar to the first category (where the organisation prioritises innovation), what assurances are there that security is embedded within the development lifecycle (DevSecOps) and that there is a culture of security-by-design?

What validation takes place and by whom? Is there segregation of duties to support testing and validation of security controls? Plus, is there an adequate level of support for ongoing maintenance of the solutions created? How are key person risks avoided?

SUMMARY

By mapping the organisation's IT environment against these categories, you can identify areas of high risk, potential technical debts, and operational dependencies. For instance:

- Centralised, tightly managed environments often exhibit fewer vulnerabilities and greater resilience due to stringent control measures.
- Dynamic, agile environments may enhance innovation, but require robust governance to prevent security lapses.

How does the organisation support how it wants to utilise IT? Are there formally agreed processes and embedded culture to manage IT within the organisation's risk appetite? And how dependent is the delivery of IT on third parties?

The categories applied to the IT environment will also impact business resilience strategies: how resilient is the environment in terms of loss of key personnel (skills and knowledge, including system topology and integration or configurations of systems) or loss of third-party providers? This step in risk profiling helps us to pinpoint key dependencies and culture so we can explore the associated operational risks further.

"The only truly secure system is one that is powered off… and even then I have my doubts."

Gene Spafford, Professor of Computer
Science at Purdue University

CHAPTER 5
HUMAN FACTORS

People are often referred to as the "weakest link" in cybersecurity, not because they always intentionally go out of their way to cause incidents, but because human behaviour is complex, inconsistent, and often unpredictable. It is precisely this complexity that makes human factors so crucial to an organisation's risk profile.

Human risk is not a single category—it encompasses individual decisions, systemic habits, cultural norms, and the motivations or vulnerabilities of those who interact with your systems, both internally and externally.

When developing a cybersecurity risk profile, failing to consider the human dimension (who your users are, how they behave, and what influences them) can leave critical blind spots in your defensive posture.

EXTERNAL HUMAN FACTORS

External human factors refer to the people who interact with your organisation from the outside: customers, partners, suppliers, or the public. Their behaviour can influence your cyber risk in unexpected ways.

A structured way to evaluate this risk is to map:

- Types of external users
- Demographic and digital capability
- Nature of interaction
- Vulnerabilities and behaviours.

Type	Demographic and digital capability	Nature of interaction	Vulnerabilities and behaviours
General public	Young adults through to senior citizens	Via website, app, email	Lack of awareness, poor password hygiene, phishing susceptibility
Customers	Varies (individuals/businesses)	Accounts, portals, e-commerce	Reused passwords, unsafe browsers, social engineering
Third-party vendors	SMEs with limited IT budgets	API integration, shared platforms	Poor cyber hygiene, outdated systems, lack of patching
Contractors	Freelancers and temporary staff	VPNs, shared drives, email	Use of personal devices, mixed loyalty, shadow IT

Using this method we can determine how an organisation, or stake-holders, may be exposed. For example, in a notable incident involving a major Australian super fund, attackers successfully accessed multiple customer accounts through credential stuffing. This technique exploits the widespread practice of [customers] reusing passwords: the attackers took credentials (usernames and passwords) stolen from breaches on unrelated platforms and systematically tried them on the super fund's login portal.

While the organisation's own security infrastructure remained uncom-promised by traditional hacking methods, this breach underscored a critical vulnerability: the human factor in cybersecurity.

The reputational fallout was immediate. Customers, alarmed by unauthorised activity on their accounts, questioned the trustworthiness of the fund despite the technical reality that the breach stemmed from user behaviour rather than a direct failure of the company's defences. This highlights a persistent challenge in digital security: even the most robust systems can be undermined by poor password practices and a lack of user awareness.

For organisations facing similar risks, several proactive measures can make a tangible difference:

- Default multi-factor authentication (MFA): requiring MFA for all logins significantly reduces the effectiveness of credential stuffing, as stolen passwords alone become insufficient for access
- User education campaigns: regularly informing and reminding users (especially customers and third parties) about the importance of unique, strong passwords and the dangers of reusing credentials across platforms
- Breach monitoring and prompt notification: implementing systems to monitor for suspicious login attempts and alerting users if their credentials appear in known data breaches help mitigate risk before attackers can exploit the breach
- Steps to limit reputational damage: transparent communication with stakeholders, swift incident response, and public demonstration of security improvements help restore trust in the aftermath of such events

Ultimately, the Australian super fund case serves as a vivid reminder that cybersecurity is a shared responsibility. Organisational controls, technical safeguards, and user vigilance must all work hand in hand to protect sensitive information and maintain trust in an increasingly interconnected digital landscape.

INSIDER THREAT

Insider threats represent a complex and evolving risk landscape for organisations, arising not only from malicious intent but also from ordinary user actions—whether careless, uninformed, or simply overwhelmed. The diversity of roles within a company means that user behaviour, and thus vulnerability to cyber threats, can vary widely depending on job function, authority, and even experience level. Recognising these nuances is crucial for crafting effective security strategies.

The following table illustrates how risk behaviours are often shaped by the unique responsibilities, routines, and pressures associated with each organisational role. Recognising these patterns helps organisations tailor their defences, training, and oversight to the specific challenges posed by executive leadership, administrative staff, developers, engineers, and others.

Role type	Risk behaviour examples
Executive leadership	Avoiding the use of MFA, over-reliance on assistants, sharing passwords, minimal security training (attendance priority)
Administrative staff	Falling for phishing, reusing passwords, clicking unknown attachments
Developers	Hardcoding credentials, skipping code reviews, bypassing secure pipelines
Engineers	Innovative and research-based increased use of shadow IT, duplicate/multiple apps
IT teams	Poor configuration management, excessive privileges, burnout
New starters	Unfamiliarity with company policies or tools

Behaviours of concern may include, but are not limited to:

- Appearing intoxicated or affected by a substance at work

- Increased nervousness or anxiety
- Decline in work performance
- Extreme and persistent interpersonal difficulties
- Statements demonstrating bitterness or resentment
- Creditors calling at work
- Sudden and unexplained wealth
- Unusual interest in sensitive or classified information.

There are several proactive ways organisations can anticipate, detect, and manage behavioural risks in the workplace, particularly those that may lead to insider threats or other forms of harm. Establishing clear policies and building a culture of awareness are essential.

For instance, implementing a whistleblower policy encourages employees to safely report unethical or suspicious behaviour without fear of retaliation. Coupled with ongoing training, organisations can equip staff to identify behavioural red flags and act early to reduce potential risks.

In high-stakes environments, like those involved in national security, this is especially critical. The Australian Defence Industry Security Program (DISP) requires all providers to actively identify and manage insider threats, not just through technology, but also through people-centred vigilance.

Here's an excerpt from DISP guidance that highlights this point:

> "If you observe any of these indicators, show an interest in that person's welfare and check if everything is okay. Simply having a conversation with them can be the first step. You must also report to supervisors observed changes in a colleague to proactively avoid serious consequences that might threaten the lives of your colleagues, Defence property or national security. This is not the time to think 'She'll be right, mate,' or, 'It's un-Australian to dob in a mate'."

This message reinforces that early intervention—even just starting a conversation—can make a difference. It's not about blame; it's about care, safety, and shared responsibility.

Organisations should also consider the following actions:

- Run behavioural risk awareness training: help staff recognise signs of disengagement, stress, and suspicious conduct.
- Foster psychological safety: encourage employees to speak up without fear of judgement or reprisal.
- Promote wellbeing checks and peer support: early signs of insider risk often stem from personal or professional stress, not malice.

By embedding a culture of awareness, empathy, and accountability, organisations can manage human-centric risks more effectively and support the overall resilience of their workforce.

EVALUATING HUMAN FACTORS

Step 1: Assess relevance

Consider these cause events. Based on your knowledge of your organisation, workforce, customers, and third-party interactions, identify which are most relevant and how they might occur in your environment.

Cause event	Level 2 cause event
Inadequate resources	Skills
	Number of staff
	Loss of key staff
Criminal activity	Theft/fraud
	Damage to assets
Disgruntled employee	Unhappiness with job or organisation, possibly seeking revenge or being careless
Management/control of staff	Inadequate supervision
	Insufficient communication
	Inadequate reporting
Compromise of functions	Abuse of rights
	Forging of rights
	Denial of actions
Human error	Non-deliberate misunderstanding of requirements, misinterpretation, misdirection, or wrong action, basic errors
	Causing harm through neglect e.g., clicking on phishing links despite training
	Social engineering: employees being tricked by external attackers into divulging sensitive information or performing unauthorised actions that can harm the organisation
	Accidental disclosure/leak of sensitive data
	Mishandling of sensitive data
Compromised user	Individual unknowingly involved in breach due to compromised account / stolen credentials

. . .

Cause event	Level 2 cause event
Insider trading	Fraudulent copying of software
	Illegal trade
Fraud	Employees engaging in fraudulent activities, such as falsifying documents, creating fake accounts, or embezzling funds
Insider espionage	Employees spying on the organisation and providing confidential information to outsiders such as competitors or foreign governments
Unauthorised activity / malicious insider	Deliberate misunderstanding of requirements, misinterpretation, misdirection, or omission of action
	Unauthorised use of equipment
	Deliberate disclosure to unauthorised individuals/entity
	Stealing data / unauthorised copying
	Sabotage
	Fraudulent copying of software
	IP theft for personal gain / benefit to a competitor
	Illegal processing of data
Workplace environment	Shift patterns
	Workload

Step 2: Determine root cause and justify relevance

For each relevant event, ask: why or how could this happen here?

Explore underlying issues, such as leadership gaps, poor onboarding, or over-reliance on a single expert.

Has this already happened in your organisation or others in your industry?

Draw on internal incident reports, whistleblower disclosures, audit findings, and external case studies.

Step 3: Evaluate the risks (insider threats)

Risk event name	e.g., loss/theft of IP
Risk description	Based on the "story", what is impacted, how, and what could happen? e.g., "XYZ data is stolen and provided to a competitor by an employee. resulting in competitive disadvantage in the market and financial loss. Achieved through unauthorised copying of the data for personal gain. Caused by inadequate segregation of duties and inconsistent background screening or check-ups on staff welfare."
Scope	Does this relate to a particular asset or business process? What's the context for the risk event?

Step 4: Identify key risk indicators

Data Source	Description
Data access patterns	Unusual or unauthorised access to sensitive data, such as accessing files or databases outside of normal working hours, accessing data outside of job responsibilities, or accessing large volumes of data
Network traffic anomalies	Monitoring network traffic for anomalies such as large data transfers, attempts to access restricted areas, or connections to suspicious external IP addresses
Account activity	Monitoring for abnormal user account activities such as multiple failed login attempts, unauthorised account creations or deletions, or changes to user privileges
System and application logs	Monitoring and analysing system and application logs for any suspicious activities, including attempts to modify system configurations, unauthorised software installations, or abnormal user behaviours
User behaviour analytics	Analysing user behaviour patterns to identify deviations from normal behaviours such as a sudden increase in file access, a change in communication patterns, or unusual system usage
Privileged user monitoring	Keeping a close eye on activities of privileged users such as system administrators or executives, as their actions can have significant impacts on the organisation's security and operations
Data exfiltration attempts	Monitoring for any unusual or unauthorised attempts to transfer sensitive data outside of the organisation's network, such as large file uploads to external sources or multiple emails with attachments
Employee dissatisfaction indicators	Tracking employee satisfaction surveys, HR complaints, and other indicators of employee dissatisfaction that could potentially lead to insider threats
Policy violations	Monitoring for policy violations such as unauthorised software installations, attempts to bypass security controls, or sharing of login credentials
Employee activity before resignation or termination	Monitoring the activities of employees who have recently resigned or been terminated to identify any abnormal or potentially malicious behaviours before their departure

Step 5: Bring it together

Describe the relevant cause event (based on scope and risk)	How plausible is the cause event? Have there been any relevant past incidents?	What controls are already in place to prevent/detect this event?	Controls we need (what can be implemented to anticipate/prevent/detect?)
Unauthorised copying of IP	Has happened in the past, after an engineer left the organisation and it was later discovered they'd copied IP data.	Standard access controls	Data loss prevention (DLP) capabilities to prevent or detect unusual behaviour and data extraction
Users disabling security controls to WFH/remotely	Has happened in the past, during COVID. Some users are still working this way.	None	Mobile device management (MDM) solution to enforce and lockdown configurations
XYZ asset stolen / lost in transit	Event has happened at least 3 times in the past 5 years. A laptop was stolen from a car.	All devices are protected and can be remotely wiped.	MDM and endpoint protection capabilities Enhanced travel awareness training

In conclusion, human behaviour, whether intentional or accidental, is a central component of cyber risk. Yet it is often the least predictable and most overlooked element of the organisational risk landscape.

By applying structured methods to identify, evaluate, and address human-related risks, organisations gain clearer insight of their exposure. This process highlights not only where controls are needed, but also where cultural and procedural improvements can significantly reduce vulnerabilities.

Ultimately, a mature risk profile isn't just about identifying threats: it's about recognising how and why they may arise, especially from the people within or connected to your organisation. Incorporating human factors strengthens your ability to build realistic, evidence-based, and resilient strategies for risk reduction.

Technology doesn't just fail on its own. Systems don't compromise themselves. Behind most cyber incidents, there is a human element: an oversight, omission, misplaced trust, poor decision or deliberate intent. If we want to get risk profiling right, we need to start by understanding people.

"The human factor is the weakest link."

Kevin Mitnick, security expert and author

CHAPTER 6
EXTERNAL DEPENDENCIES

Every organisation, regardless of sector or scale, is intrinsically linked to a network of external partners. These third-party relationships are more than just operational conveniences—they are vital threads woven into the fabric of an organisation's day-to-day functioning and long-term resilience. However, these connections also introduce unique risks, especially when they pertain to essential services, data management, or core business activities.

Effectively managing these external dependencies starts with gaining a clear, unified picture—best achieved through the development of a comprehensive third-party register. This register should serve as a dynamic resource, detailing not only who your external suppliers and service providers are, but also the nature and criticality of the services they deliver, their contract status, and any associated risks.

Building and maintaining such a register are collaborative tasks that draw on the expertise of multiple departments. Finance contributes valuable insights through payment records and spending analyses; procurement provides access to contract details and expiration timelines; IT manages information about technology vendors and service agreements, while operational risk and business continuity teams help prioritise dependencies according to their impacts on critical functions.

By synthesising data from across these stakeholders, organisations can create a fuller, more nuanced map of their external landscape. This collective approach not only sharpens visibility over third-party risks but also strengthens the organisation's ability to respond to disruptions, negotiate effectively, and uphold continuity even when external events threaten to destabilise operations.

FINANCE

- Obtain payment records for services, including software licences and subscriptions.
- Cross-check recurring payments to identify long-term, critical providers.

PROCUREMENT

- Access the contract register, which should detail providers, their roles, and the services they deliver.
- Identify providers with expired or soon-to-expire contracts and assess the risks of disruption.

IT

- Review the IT asset registers for licensed software, managed services, or IT-specific vendors.
- Evaluate SLAs and support contracts for critical systems.

OPERATIONAL RISK (BUSINESS CONTINUITY)

- Leverage the BIA for insights into dependencies critical to operational resilience.
- Identify providers linked to high-priority business functions.

A comprehensive third-party register must include specific data points to effectively evaluate risk exposure and dependency criticality. These elements provide the necessary context for risk-based decision-making and due diligence prioritisation.

Beyond basic provider identification, the register should capture detailed information about service scope, access levels, and potential impacts of disruption. This multidimensional view allows for more nuanced risk assessment and more targeted mitigation strategies.

The third-party register should cover:

Provider information	Name, contact details, and primary point of contact
Services/products provided	A description of what the provider delivers
Dependency scope	The departments or operations relying on this provider
Access levels	Logical (e.g., data systems) and physical (e.g., onsite access)
Data and asset access	Whether the provider has access to sensitive organisational data or physical assets
Impact of service loss	The operational consequences of losing the product or service
Impact of provider loss	The fallout if the provider ceased operations or experienced a disruptive event

Not all third-party relationships carry the same level of risk. Classifying providers based on their criticality and risk profile allows for proportionate due diligence and oversight. This targeted approach optimises resource allocation while ensuring appropriate controls for high-risk relationships.

Critical providers warrant enhanced scrutiny, including validation of specific controls and supply chain analysis. Standard security questionnaires are insufficient. These relationships require deeper examination of specific risks, validation of controls, and consideration of extended supply chain dependencies.

- Go beyond standard (and very basic) security questionnaires, explore the key risks associated with the products and services provided, and obtain validation that the specific controls are adequate.
- Consider identifying their providers: what does the overall supply chain look like and is there a critical partner that may introduce otherwise unknown risks?
- Validation of controls should be specific to the risks imposed and must be monitored over time and to a degree commensurate with the risk.

Standard providers with minimal impact can be managed through lighter touch processes. This risk-based classification maintains that the level of oversight matches the potential exposure.

- These types of relationships can be covered by standard third-party questionnaires and review of the data they provide in their collateral, such as within their trust centres.
- This may be a one-off review prior to procurement or subject to twice yearly updates (or when a change occurs).

Robust evaluation of third-party dependencies enhances an organisation's ability to mitigate risk and maintain resilience. By categorising and prioritising providers based on criticality and risk exposure, organisations can develop targeted strategies for managing dependencies.

Strengthening relationships with key partners, ensuring alignment on risk-management practices, and maintaining a current third-party register are vital steps in reducing vulnerability and ensuring continuity.

"Outsourcing of an activity does not relieve management and the board of their responsibility."

FFIEC IT Handbook

CHAPTER 7
INDUSTRY BENCHMARKING

Benchmarking is a powerful tool in the context of risk profiling, enabling organisations to measure their security practices, exposures, and resilience against those of their industry peers.

By systematically comparing incident histories, control effectiveness, and threat landscapes across similar organisations, benchmarking provides critical context for appreciating where your organisation stands in terms of risk.

This process not only highlights existing gaps and strengths, but also brings to light emerging risks that may not yet have been internalised. Integrating benchmarking into your risk-profiling approach allows for more nuanced, data-driven decisions, ensuring that your cybersecurity posture is informed by both internal assessment and the lived experience of the wider industry. Ultimately, this strengthens your ability to prioritise investment, develop targeted mitigation strategies, and maintain a proactive stance in the face of evolving threats.

For evaluation and to refine your industry benchmarking, consider utilising the following guiding key questions. These prompts are designed to bring you deeper insight into your organisation's relative standing and inform more effective risk-management decisions.

INCIDENT HISTORY

- Have any of your peers suffered data breaches or cybersecurity incidents?
- If so, what were the contributing factors (e.g., human error, software vulnerabilities, weak controls)?
- How did they respond and what were the outcomes (e.g., regulatory fines, reputational damage)?

TECHNOLOGY STACK

- Do they use the same or similar hardware, software, or third-party services?
- Are there shared vulnerabilities in widely used tools or platforms in your industry?

SHARED THREATS

- Are your peers facing similar threat actors, tactics, or attack vectors (e.g., ransomware, phishing, supply chain attacks)?

MATURITY LEVEL

- How does your organisation's cybersecurity maturity compare to others in your sector?
  - Investigate how competitors advertise their cybersecurity efforts. Are they highlighting certifications, partnerships, or advanced security technologies?
  - Use their public messaging to identify gaps or opportunities in your own cybersecurity posture.
- Are they adopting advanced technologies (e.g., AI for threat detection, zero trust architectures)?

There are various avenues to research and keep up to date with changes to the industry's risk profile. Participate in industry forums and special interest groups to gain insights into shared threats and industry standards.

Set time aside regularly to research and leverage publicly available data through news channels and data breach reports to learn about wider threats and common pitfalls.

Stay informed through CTI feeds to identify emerging threats and shared vulnerabilities in your industry.

Benchmarking against peers and learning from their experiences are essential components of cybersecurity risk management and refining your own organisation's risk profile. By leveraging industry insights, standardised frameworks, and advanced tools, organisations can enhance their defences, address vulnerabilities proactively, and maintain a competitive edge in their cybersecurity maturity.

"If you know the enemy and know yourself, you need not fear the result of a hundred battles."

Sun Tzu, Military Leader & Strategist

CHAPTER 8

UNDERSTANDING STAKEHOLDER NEEDS AND EXPECTATIONS

Understanding stakeholder needs and expectations is a fundamental component of effective risk profiling, as it provides a holistic view of the organisation's vulnerabilities and priorities; hence it is one of the first clauses covered in ISO 27001.

Stakeholders—ranging from executives and employees to clients, regulators, and third parties—bring unique perspectives and expectations that shape the organisation's risk landscape.

By engaging with stakeholders, an organisation can identify critical assets, clarify the impacts of potential risks, and verify that cybersecurity strategies are aligned with broader business objectives. This alignment not only enhances the accuracy of the risk profile, but also fosters buy-in and collaboration, making it easier to prioritise resources, address vulnerabilities, and build resilience against evolving threats. Ultimately, incorporating stakeholder perspectives confirms that risk profiling is not just a technical exercise but a strategic tool for safeguarding organisational success.

Stakeholders can include individuals, teams, departments, external groups, clients, regulators, and partners—all of whom may have unique expectations around how cybersecurity supports organisational objectives.

RESEARCH

Begin by listing all internal and external stakeholders based on your grasp of the organisation's structure, objectives, and operations. Typical categories include:

- Executive leadership
- Department heads (e.g., HR, finance, IT)
- Operational teams
- Customers and clients
- Regulators and auditors
- Vendors and third-party providers

Engage with key personnel across the organisation to ensure your list is comprehensive. Ask questions to uncover potential overlooked stakeholders, including informal influencers or peripheral groups that interact with the organisation's operations.

Stakeholder (or "interested party") expectations around information security will vary greatly, so it's very important to understand their unique perspectives and needs so you can verify whether the current approach, objectives, and outcomes of the security strategy will meet these needs.

For instance:

- Executives may prioritise risk reduction and regulatory compliance.
- IT teams may focus on minimising disruption and maintaining system integrity.
- Clients may expect data confidentiality and swift issue resolution.
- Regulators require adherence to specific security standards or frameworks.

This exercise guarantees that the cybersecurity objectives align with the organisation's objectives and values, and are therefore easily absorbed and accepted across the organisation.

Conduct this exercise frequently to verify that the cybersecurity objectives and key results (OKRs) remain in line with the organisation and its stakeholders' needs and expectations. Conduct further research over time to fully understand where cybersecurity fits and benefits others.

Here are some activities you can consider in investigating:

1. **Conduct stakeholder interviews or workshops**

- Hold one-on-one interviews or group workshops with representatives from each stakeholder group.
- Use open-ended questions to explore their concerns, priorities, and expectations around cybersecurity.

2. **Map stakeholder needs**

- Create a matrix or table outlining each stakeholder group, their specific expectations, and how these align with current security objectives.

Stakeholder group	Key needs/expectations	Current alignment	Gaps identified
Executive team	Compliance with regulations	Partially aligned	Need more reporting transparency
Clients	Data protection and responsiveness	Fully aligned	None

3. **Feedback mechanisms**

- Distribute questionnaires or surveys to gather structured feedback from a wider audience.
- Include questions on their perceptions of security effectiveness, areas for improvement, and communication preferences.

4. **Shadow stakeholders**

- Spend time observing how different departments operate. This helps identify implicit expectations or pain points that stakeholders may not articulate or even be aware of.

The end goal is to implement cybersecurity objectives that support organisational objectives and stakeholder values.

For example:

- If the organisation's core mission emphasises customer trust, cybersecurity must focus on data protection and swift breach response.
- If operational efficiency is a key objective, the security strategy should minimise disruption caused by security controls.

Frequent alignment checks are critical, especially as organisational priorities, market conditions, and regulatory landscapes evolve.

COMMUNICATE

Create a communications plan that facilitates regular engagement with the stakeholders and proactively demonstrate how the cybersecurity strategy is enabling the organisation and meeting their needs and expectations.

Create a plan that:

- Enhances user engagement and feedback on a regular basis
- Acknowledges their concerns, needs, and expectations
- Builds relationships across the organisation
- Actively communicates the progress of cybersecurity initiatives
- Shares success stories
- Delivers the cybersecurity strategy and its objectives over time.

Acknowledging and aligning with stakeholder needs are essential for embedding cybersecurity within the organisation. Without this alignment, security efforts are in danger of becoming siloed and disconnected from broader business objectives. Through actively engaging stakeholders, gathering feedback, and adapting to their expectations, you can build a security strategy that is not only effective, but also supported and even embraced across the organisation.

"If you think technology can solve your security problems, then you don't understand the problems and you don't understand the technology."

Bruce Schneier, cybersecurity expert, cryptographer, author, and public speaker

CHAPTER 9
ADOPTION OF ARTIFICIAL INTELLIGENCE

Digital transformation is an ongoing reality shaping how organisations operate, compete, and innovate. As businesses embrace technologies like AI, they also step into a new era of uncertainty and evolving risk exposure. These technologies bring undeniable opportunities for efficiency, agility, and innovation, but they also demand shifts in how we think about risk profiling, governance, and resilience.

This chapter explores how the adoption of AI will reshape an organisation's risk profile and why it's essential to continually review and update the organisation's risk-evaluation criteria and toolkit. As technology changes, so too must our approaches to risk governance, stakeholder engagement, and control validation.

AI STRATEGY

Many organisations are embracing the possibilities that AI can bring and it's an exciting opportunity for CIOs and executives, but let's not forget that it's a tool and—just like every other technical change—its utilisation needs to be considered and deliberate.

While the adoption of AI offers immense opportunity, diving in without a deliberate and well-structured approach exposes organisations to significant and sometimes unforeseen risks. Many businesses are eager to experiment with AI, encouraging innovation and rapid adoption, but this enthusiasm can often lead to fragmented efforts and a lack of cohesion between AI initiatives and overall business objectives. AI affects decision-making, ethics, and organisational trust in ways that legacy systems never did.

Ethical lapses in AI use, such as use of biased recruitment models or surveillance misuse, can trigger reputational crises that outpace even the technical failures. Ethical governance must therefore be considered part of the organisation's strategic risk posture.

Strategic planning compels organisations to pause and ask these important questions: Why do we want to adopt AI? What specific benefits are we seeking? How will we measure success? And which tools align best with our goals?

By clearly defining the purpose and intended outcomes, organisations can verify that AI adoption is intentional, aligned with business strategy, and responsive to genuine needs—rather than simply chasing trends or technological novelty.

For instance, many organisations may feel enthusiastic about integrating AI into their operations, recognising the importance of providing staff with guidance and training on its responsible use. However, to fully realise the benefits and mitigate the risks, it is crucial to invest in equipping staff with prompt-engineering skills and deeper grasp of how to supply advanced context when interacting with AI tools. By training employees not only in acceptable use but also in crafting precise prompts and leveraging contextual information, organisations empower their teams to achieve accurate, reliable, and value-aligned outcomes from AI systems.

Yet, without first establishing which AI tools are appropriate, clarifying acceptable-use standards, and identifying the objectives driving AI adoption, efforts such as policy development, technical training, or prompt-engineering workshops can risk misalignment with broader organisational goals.

This common scenario highlights the necessity of building a solid strategic foundation prior to embarking on AI initiatives. By establishing clear objectives, usage guidelines, and comprehensive training (including advanced prompt engineering) at the outset, organisations can ensure that subsequent governance, integration, and continual skill development remain relevant, effective, and resilient to future changes in technology or business priorities.

It's critical for all sizes of organisations to provide all staff with guardrails that reinforces the adoption of AI in line with the organisation's risk appetite.

Moreover, strategic planning enables a proactive approach to risk management. AI doesn't necessarily create entirely new risks, but it does change the speed, scale, and nature of existing ones. Decision-making failures, ethical considerations, and data breaches can occur more quickly and with broader impact. Without a clear strategy, organisations risk opening a Pandora's box of governance challenges that are far more difficult and costly to manage after the fact.

In short, strategic AI planning bridges the gap between innovation and organisational resilience. It guarantees that the benefits of AI are harnessed safely, sustainably, and in alignment with the business, rather than introducing new vulnerabilities or undermining stakeholder trust. An AI risk-profiling checklist (provided in Appendix R), for example, can be a powerful tool in assessing readiness and guiding responsible adoption across the organisation.

ALIGNING AI INITIATIVES WITH ORGANISATIONAL STRATEGY: FOUNDATIONS FOR SUCCESS

As organisations accelerate their journeys into the realm of AI, the necessity for strategic clarity and operational discipline becomes more pronounced than ever. The traditional approach of setting a course and following it strictly simply doesn't hold up in the shifting terrain that AI presents. With the speed at which AI technologies evolve, a static strategy becomes obsolete almost as soon as it is penned.

To truly harness the transformative power of AI, organisations must shift from rigid, long-term roadmaps to strategies that are dynamic, responsive, and iterative. Instead of isolated proof-of-concept projects, successful organisations ground their AI initiatives in frameworks that can flex and adapt as both business needs and technology landscapes change. This is not just an operational necessity, but a strategic imperative: only by embedding adaptability into the core of their AI strategies can organisations remain competitive and resilient.

KEY PRINCIPLES FOR ADAPTIVE AI STRATEGY

- Strategic alignment: every AI initiative is clearly linked to evolving business priorities and continually reassess this alignment as and when circumstances change.
- Outcome-focused planning: define success in terms that allow for change, balancing the pursuit of innovation with rigorous risk management and organisational goals.
- Integrated, flexible frameworks: move beyond siloed planning to weave AI strategies into broader digital-transformation efforts, business continuity plans, and ongoing risk assessments, so that governance and innovation develop hand in hand.
- Stakeholder-centric value: demonstrate clear, adaptable value propositions to customers, employees, and partners, recognising that what stakeholders need today may change tomorrow.

OPERATIONALISING ADAPTABILITY

The most effective organisations do not attempt to predict exactly which AI tools or applications they will use years in advance. Instead, they establish frameworks that support frequent strategic reviews, modular policies that can be rapidly updated as technology evolves, and robust feedback loops that confirm strategies remain relevant.

In this fast-changing environment, adaptability is the cornerstone of resilience. Strategy has become a living process—one that must be revisited, revised, and reimagined with every new development in AI.

By prioritising agility and continual alignment, organisations position themselves to navigate emerging risks and seize opportunities, converting the volatility and uncertainty of the AI era into a competitive advantage.

"There is no question that artificial intelligence
needs to be regulated. It is too important not to."

Sundar Pichai, CEO of Google and Alphabet Inc

CHAPTER 10
AI GOVERNANCE

In 2025, AI governance stands as one of the most pressing and complex risk-management challenges for organisations. Unlike traditional technologies, AI systems introduce unprecedented risks, ranging from a lack of transparency (explainability) and algorithmic bias to complex data-governance concerns and rapidly evolving regulatory landscapes. These factors have the potential to profoundly affect not only daily operations, but also the long-term trust of stakeholders and the reputation of the organisation.

AI governance is much more than a compliance checkbox—it is a strategic discipline that guides how an organisation manages AI throughout its entire lifecycle, from initial design and development, through deployment and ongoing operation, all the way to decommissioning. Frameworks such as the ISO 42001 Artificial Intelligence Management System (AIMS) provide much-needed structure, ensuring that AI deployments are carefully defined, tightly controlled, and do not inadvertently expose the organisation to unnecessary risk.

However, effective AI governance should extend beyond mere regulatory adherence. When approached strategically, it becomes a powerful enabler—empowering organisations to realise AI's transformative potential, all while bolstering operational resilience and fostering enduring confidence among customers, partners, and broader stake-holders.

UNDERSTANDING THE EVOLVING THREAT LANDSCAPE

Before implementing robust governance frameworks, organisations must fully grasp the dynamic and rapidly changing threat landscape introduced by AI. Some of the most critical challenges are:

- Rising security breaches: the average cost of an AI-related security breach has climbed to US$4.8 million globally, reflecting the increasingly sophisticated nature of attacks targeting AI systems.
- Pervasive compliance concerns: 67% of organisations now report significant anxiety around compliance and regulatory obligations specific to AI, highlighting widespread recognition of the risks involved.
- Escalating penalties: regulatory fines tied to AI governance failures are increasing year on year, with a marked 40% annual rise as new laws and frameworks come into force across different jurisdictions.
- Governance failures undercut success: alarmingly, 45% of AI projects fail to deliver their intended outcomes due to the absence of sufficient governance structures and oversight.

Emerging threat vectors

The complexity of AI governance is heightened by a range of emerging threats, including:

- AI-driven cyber attacks: threat actors now employ AI to automate and personalise phishing campaigns, leveraging deepfake technology and advanced social-engineering techniques to deceive employees and compromise systems.
- Model manipulation: malicious actors exploit vulnerabilities in AI models, executing adversarial attacks such as data

poisoning to skew AI decision-making or undermine trust in automated outputs.

- Regulatory compliance complexity: with AI governance requirements evolving rapidly and differing across countries, organisations face mounting challenges in ensuring consistent compliance and managing cross-border data flows.
- Algorithmic bias and discrimination: without vigilant oversight, AI systems can entrench or amplify biases, resulting in unfair or discriminatory outcomes that may expose the organisation to legal action and reputational harm.

By proactively addressing these risks through comprehensive governance frameworks and continuous oversight, organisations not only reduce exposure to costly incidents, but also build a foundation for sustainable AI adoption that aligns with business goals and societal expectations.

STRATEGIC AI GOVERNANCE FRAMEWORK

AI governance helps refine how and why AI is to be managed throughout its lifecycle, from design, development, and operation to end of life. ISO 42001 AIMS has been designed to facilitate AI governance, ensuring that its use is defined and controlled, and doesn't expose the organisation to unnecessary risks.

Before rolling out AI, organisations should:

1) Define the strategic objectives.

2) Provide clear policy statements that:

- Define which AI systems are authorised
- Outline the acceptable use of AI
- Outline how AI-specific risks are to be managed
- Define clear roles and responsibilities that cover the full system lifecycle.

3) Identify the regulatory frameworks relevant to the use and security of AI systems (such as ISO 27001, ISO 42001, Australian Privacy Act).

4) Define how AI risks are to be managed and provide supporting procedures such as AI impact assessments, (AIIAs) and update any risk-evaluation tools (e.g., risk-cause event library, controls library, risk appetite).

5) Evaluate cultural readiness for the adoption of AI. How can the organisation foster a workforce that feels confident in challenging AI decisions, recognise the boundaries of automation, and is trained to identify misuse or drift?

Control and innovation in AI governance

Research indicates that restrictive AI policies may lead to unintended risks within organisations. When AI tools are broadly prohibited, employees might continue using them outside of official channels; for example:

- Accessing personal devices not subject to corporate oversight or DLP
- Using consumer AI platforms (such as ChatGPT, Claude, Gemini) with personal accounts
- Engaging with cloud services that bypass organisational firewalls
- Utilising mobile applications beyond the scope of IT monitoring.

Such practices can increase the likelihood of shadow IT, which may result in:

- Reduced capacity to maintain necessary logs for detecting data breaches
- Limited oversight and control over data classification and handling
- Decreased visibility into user behaviours.

The purpose of AI governance is to facilitate safe, effective, and compliant AI adoption aligned with organisational objectives. Recognising that outright prohibition may elevate risk rather than mitigating it, organisations can consider approaches that:

- Address employee preferences while establishing appropriate controls
- Support innovation within defined risk parameters
- Promote transparency in governance processes
- Seek competitive advantage while safeguarding assets.

Effective AI governance programs often factor in employee needs, preferences, and behaviours alongside technical measures. By aligning frameworks with these factors, organisations reduce unmanaged AI use and support responsible innovation.

Ultimately, the aim is not absolute control, but a balanced approach to risk management that allows for the productive use of AI technologies while maintaining security and compliance.

Roles and responsibilities

Change in the use of technology may mean you need to engage new stakeholders into the risk-governance process such as data scientists or ethics boards. To identify who needs to be involved, consider:

- Which departments "own" AI-enabled systems
- Who needs to be involved in the decision processes or risk acceptance (Consider: CIO, risk officer, general counsel, data science lead)
- Whether new roles are required e.g., chief AI officer, AI risk committee
- What policies are needed to embed explainability and accountability
- Who is responsible for escalating AI system errors or performance degradation
- How AI risks are communicated to the board or external regulators.

Boards must be equipped with high-level summaries of AI-related risk exposures, supported by metrics that reflect the AI's performance and ethical use. This could include fairness scores, error rates, or incident logs translated into business risk terms to maintain appropriate oversight at board level.

AI RISK-MANAGEMENT TOOLS

Before you embark on risk evaluation, it is essential to validate that the tools you are using to identify and evaluate risks are up to the task. AI may introduce new categories of risk that have not been considered in the past, such as the following.

Risks *to* the model:

- Data or model drift: could data or performance degrade silently over time?
- Data quality: how could corrupted, incomplete, or out-of-date data be identified?
- LLM manipulation: could the model itself be compromised (e.g., prompt injection, spoofing)?
- LLM poisoning: how to deal with this sophisticated cyberattack that targets the training data or fine-tuning process of LLMs, representing one of the most insidious security risks in AI systems today?

Risks *because* of the model:

- Hallucinations: how to identify information that sounds plausible but is factually incorrect, made up, or unsupported by real data?
- Automated decisions: what are the implications of automated decisions made without human oversight?
  - How might bias or poor data quality in AI models create downstream risks?
  - How to identify inappropriate or unsafe recommendations?
  - How to eliminate misinformation?

As generative and autonomous AI tools become embedded into day-to-day operations from marketing to finance, there is growing concern over systems making decisions without sufficient human validation. This increases the risks of unintended consequences, reputational damage, and compliance violations.

- Are legacy systems compatible with AI tools? What interoperability challenges exist?
- Is there a lack of transparency or explainability?
- Are there human interaction risks: blind trust in AI outputs, human error?

For existing business processes, how will the AI tooling reshape the risk profile? If an organisation is moving away from human data analysis, for example, what happens when AI models are trained on flawed or incomplete data?

These questions should trigger a review of the risk-cause register: are there new avenues we must consider that could cause such events, such as using unvalidated or unvetted data?

Additionally, we may need to consider new information-security controls beyond ISO 27001 standard control requirements. Although existing controls, such as access management, can limit access to the model and backend data, the necessary depth of the controls and other measures such as data governance are required to validate the integrity of the system and confirm its data can be trusted.

AI IMPACT ASSESSMENTS

AIIAs are essential in helping us to refine the purpose of AI systems and understand the associated risks, especially within highly regulated industries. An AIIA must consider the full AI system lifecycle, as risks can occur at any stage.

The AIIA should clearly define:

- The asst owner: who is authorised to define its purpose and objectives, and to accept the risks associated with its use
- The intended purpose and intended result of the AI system
- Business processes that will be supported by the new AI system
- Intended target user groups
- Roles and responsibilities
- The type and classification of data authorised during the lifecycle:
  - What data is authorised for testing?
  - What data will be processed in the production environment?
  - How will data quality be safeguarded?
- In using the AI system, what aspects are explainable:
  - Are technical processes and human decision points clear?
  - Are the technical components understood and transparent?
  - Can you explain how and why the system makes specific decisions?
- The relevant standards and regulations applicable to the supported processes and data. Note: consider additional requirements where personal data may be processed by the AI system (e.g., following a privacy impact assessment)
- Risks associated with the model covering design, development, production, and end of life. Risks identified should specify:
  - Potential cause events
  - Impacts of loss of confidentiality, integrity, and availability at each stage of its lifecycle
  - The baseline controls assumed to be in place
  - Specific controls required to mitigate associated risks.
- Third-party and supplier risks. Consider:
  - Lack of transparency in third-party models
  - Dependency on external AI ethics or control practices
  - Supply chain risks if vendors don't adhere to security or fairness standards.

Where AI services are delivered via third-party platforms, organisations must treat those providers as critical suppliers. Procurement and risk-management processes must evolve to include AI-specific due diligence, such as reviewing model transparency, data lineage, and vendor governance maturity.

What metrics or KPIs can be used to track AI-related risks (e.g., model error rates, fairness indicators, unplanned overrides)? See the example AI risk and performance metrics table in Appendix S.

- How can feedback loops be built into the AI lifecycle to capture unintended consequences?

Define the process should the AI system fail or behave erratically or become corrupted. Identify whether AI-specific response and recovery plans are required.

- How are risks to be handled prior to launching the AI system? Which risks may still be evident and move into production that will need to be accepted and managed by the asset owner?

In summary, we shouldn't treat AI systems any differently to any other technical change; by following structured governance and risk-management practices, organisations can harness AI responsibly, avoiding new vulnerabilities while supporting innovation and stakeholder trust.

"In the digital age, data privacy and security are not just compliance requirements; they are fundamental to maintaining customer trust."

Pierre Nanterme, former Chairman and CEO of Accenture

CHAPTER 11
REGULATORY COMPLIANCE

Regulatory compliance plays an important part in an organisation's risk profile as it helps set the framework and benchmark for what is expected within the respective industry. Much like other regulated disciplines, cybersecurity cuts across the organisation and in recent years we've seen a significant global uplift in regulations and standards that incorporate information security, cybersecurity, privacy, and AI governance which must be captured when defining the context for an organisation.

The regulatory landscape has become increasingly complex, with organisations operating across multiple jurisdictions and facing overlapping and sometimes conflicting requirements. This complexity is amplified by the rapid emergence of AI-specific regulations that require new governance approaches beyond traditional cybersecurity frameworks.

It's important to engage and collaborate with your legal and compliance team to clearly define what's expected across all jurisdictions where your organisation operates. As with stakeholder needs and expectations, it is essential to fully define the regulatory requirements, what they apply to, and how the requirements are met through your integrated management systems.

For further detail, please refer to the tables in Appendix G that summarise current global regulations and standards. These resources provide a clear overview of key requirements across major jurisdictions, serving as a valuable reference as you navigate compliance and governance challenges.

INFORMATION AUDIT

Privacy and data-protection legislation has a strong tie to information and cybersecurity, as organisations must be able to demonstrate the secure handling of personal data through its entire lifecycle. Privacy legislation worldwide has evolved significantly in recent years, with regulations like GDPR, CCPA, and others placing strict requirements on data handlers. These frameworks share common principles: transparency about collection practices, purpose limitation, data minimisation, and security requirements.

It is essential that, if your organisation processes personal data, you conduct a full audit of what information is collected, where from, how, what happens to it, who has access, who it is shared with (and why), where is it stored, for how long and how it is disposed of.

Examining this lifecycle enables security professionals to identify vulnerabilities at transition points where data moves between systems, departments, or external entities. These transition points often represent the greatest risk for unauthorised access or data leakage.

A thorough information audit requires cross-departmental collaboration, as data often traverses multiple business functions. Begin by establishing a dedicated audit team with representatives from IT, legal, operations, and business units who handle personal information.

Effective audits employ multiple discovery methods, including system scans, stakeholder interviews, documentation review, and process observation. The goal is to create a comprehensive inventory that will serve as the foundation for compliance efforts and security enhancements.

Your audit should examine both the formal data-management policies and the actual practices occurring within the organisation. This gap analysis often reveals informal processes or shadow IT systems that operate outside governance structures.

Pay special attention to third-party relationships where personal data is shared externally. These relationships require proper contractual protections and ongoing oversight to confirm partners maintain appropriate security standards.

Common vulnerability points:

Lifecycle stage	Common vulnerabilities	Mitigation approach
Collection	Excessive data gathering, unclear consent	Data minimisation, transparent notices
Storage	Unencrypted databases, excessive access	Encryption, access controls, segregation
Processing	Purpose creep, unauthorised analytics	Purpose-limitation policies, monitoring
Sharing	Insecure transfer methods, inadequate contracts	Secure protocols, vendor assessments
Disposal	Incomplete deletion, lack of verification	Certified destruction, disposal confirmation

Information audits frequently uncover similar vulnerability patterns across organisations. The transition points between lifecycle stages present particular risk, as do interactions with third-party vendors and cloud services, where visibility may be limited.

Legacy systems often present significant challenges, as they may lack modern security features but still contain valuable historical data. These systems require special attention and compensating controls when they cannot be easily upgraded or decommissioned.

BENEFITS BEYOND COMPLIANCE

An information audit delivers value beyond regulatory compliance by providing a comprehensive view of your data ecosystem. This visibility enables better resource allocation, improved risk management, and more effective security investment targeted at your most valuable information assets. For example:

- Comprehensive data mapping eliminates redundant storage and streamlines information flows, reducing overhead costs and maintenance requirements. Organisations often discover opportunities to consolidate systems and standardise data-handling practices.
- By mapping exactly what data exists and where it resides, teams can make more informed decisions about technology investments and process improvements.
- Organisations with mature data-governance practices gain competitive advantages through improved decision-making and greater agility. They can more confidently leverage data assets while maintaining appropriate protections.
- When privacy considerations are integrated into product development and business operations from the outset, organisations can innovate more freely without compliance concerns creating roadblocks.

"It is not the strongest of the species that survive, nor the most intelligent, but the one most responsive to change."

Attributed to Charles Darwin

CHAPTER 12
STRATEGIC CYBER THREAT INTELLIGENCE

Strategic CTI seeks to examine the internal and external cyber threat landscape in which an organisation operates and to identify any new or emerging risks which may impact the organisation's business mission.

Effective strategic CTI requires in-depth research on the specific threats that your industry is exposed to, for example, ransomware attacks in healthcare and financial fraud in banking, and this research utilises internal threat intelligence data, third-party feeds, and industry reports to collect intelligence.

Having established the organisation's stakeholders, we can now delve deeper into the key components of risk profiling. These components uncover the organisation's vulnerabilities, identify assets at risk, and develop probable threat scenarios aligned with its operations and objectives—also referred to as strategic CTI.

Six essential activity steps support this process:

1. PESTEL analysis
2. BIA
3. Industry profiling
4. Tactical and operational CTI
5. Control maturity assessments

6. Enhancing visibility

Each of these activities provides critical insights into the risk land-scape, together forming a comprehensive profile of the organisation's cybersecurity posture. Let's explore each in detail.

PESTEL ANALYSIS

PESTEL analysis examines external and internal influences that can impact the organisation's objectives, providing a structured approach to monitoring and prioritising risks. It highlights factors beyond the organisation's control, while identifying opportunities to strengthen the cybersecurity strategy.

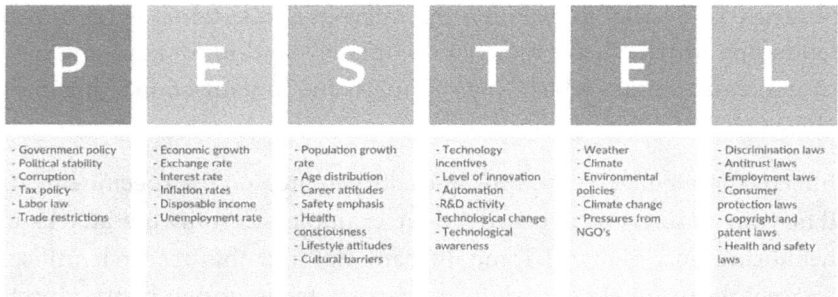

P	E	S	T	E	L
- Government policy - Political stability - Corruption - Tax policy - Labor law - Trade restrictions	- Economic growth - Exchange rate - Interest rate - Inflation rates - Disposable income - Unemployment rate	- Population growth rate - Age distribution - Career attitudes - Safety emphasis - Health consciousness - Lifestyle attitudes - Cultural barriers	- Technology incentives - Level of innovation - Automation -R&D activity Technological change - Technological awareness	- Weather - Climate - Environmental policies - Climate change - Pressures from NGO's	- Discrimination laws - Antitrust laws - Employment laws - Consumer protection laws - Copyright and patent laws - Health and safety laws

We can utilise the same methodology within cybersecurity, with a specific focus on how each PESTEL category may impact the organisation's approach to cybersecurity.

First, complete the evaluation by considering the external influences that you may have no control over:

P	• Evaluate the political landscape and its potential impact on cybersecurity (e.g., government incentives or regulations promoting enhanced cyber defences). • Look for grants, national cybersecurity programs, or initiatives to bolster organisational capability. • Assess geopolitical tensions that may heighten risks, such as cyberwarfare or sanctions.
E	• Analyse how economic changes (growth, recession, mergers, or downsizing) affect security priorities. • Plan for secure onboarding in cases of acquisition and the mitigation of insider risks during layoffs.
S	• Understand stakeholder expectations regarding cybersecurity. • Consider the reputational benefits of robust security and how it influences trust and loyalty
T	• Evaluate how technological advancements like AI, cloud adoption, or other enhancements (e.g., quantum cryptography) impact the organisation's risk profile. • Identify emerging vulnerabilities introduced by new systems or innovations.
E	• Consider location-based risks, such as natural disasters or climate-related changes, that might affect operations or infrastructure. • Prepare for regional risks if expanding operations into new geographies.
L	• Stay informed on evolving regulations and compliance requirements (e.g., GDPR, SOCI Act, Cybersecurity Act 2024). • Assess the penalties of non-compliance and the operational adjustments needed to meet legal expectations.

Beyond external factors, PESTEL analysis can also be used to evaluate internal organisational dynamics:

P	• Is cybersecurity adequately supported by leadership? Are there shifts in governance that may affect resource allocation or strategic alignment? • Is the organisation planning a change in governance structure (maybe linking to how the organisation is reacting to external changes)? How might this impact the cybersecurity strategy? Will specific roles and responsibilities change and is there a change in direction or priorities for the organisation? • What do you need to consider in ensuring the security objectives align?
E	• Are security needs competing with other financial priorities?
S	• What are the social implications of doing security well? What are the stakeholder needs and expectations, and how might they influence your strategy?
T	• Aligned with the changes to organisational strategy, growth may introduce new complexities to the IT environment or new systems; what assurances are needed to manage this change? • Will changes in strategy or operations (growth, downsizing, or IT restructuring) impact the organisation's technical risk profile? (as outlined in Chapter 2)
E	• How might sustainability initiatives (e.g., reducing the carbon footprint) influence technology use and risks?
L	• What impact do new/updated regulations have on your organisation and approach to cybersecurity? What are the implications/complexities in complying? What are the implications of non-compliance?

BUSINESS IMPACT ANALYSIS

BIA stands as a foundational pillar in identifying and managing organisational risk, beyond its traditional role as merely a prerequisite for disaster-recovery planning. An effective BIA shines a light on the intricate web of interdependencies within the organisation—across people, processes, technology, and third parties—providing a holistic perspective on what truly drives operational resilience.

Through deep and systematic BIA, organisations gain the insight necessary to identify and prioritise critical business functions, reveal potential single points of failure, and clarify which assets or processes are most vital to continued success. This clarity allows business leaders to move beyond intuition, enabling data-driven decisions about risk tolerance, investment, and resource allocation.

Crucially, the relevance of BIA to risk profiling cannot be overstated. The outputs of a robust BIA directly inform risk assessment by mapping the cascading impacts that disruptions in one area may have across the organisation. By linking business functions to their supporting assets—such as IT systems, facilities, and personnel—a BIA provides the empirical foundation needed to assign meaningful risk profiles across domains including ICT, supply chain, legal and regulatory compliance, and environmental considerations.

Rather than being relegated to IT or continuity teams, a well-executed BIA should serve as a cross-functional resource, aligning risk priorities across the business. It facilitates the identification of high-impact vulnerabilities and the development of mitigation strategies that are both targeted and proportionate to actual business needs.

While the specific mechanics of conducting a BIA are well-documented in industry standards, this book emphasises leveraging BIA insights as a dynamic tool for operational risk management and resilience planning. Ultimately, integrating BIA findings into the organisation's risk-profiling framework ensures that security, compliance, and continuity activities are grounded in genuine appreciation of what matters most —enabling the organisation to anticipate, withstand, and recover from disruptions of all kinds.

Key outcomes

A thorough BIA report will allow you to:

1. Identify critical business functions: define which processes are vital to operations, their interdependencies, and how they impact the organisation's overall mission
2. Link supporting assets to business functions: map IT systems, personnel, facilities, and supply chains to critical processes, highlighting dependencies
3. Categorise risk profiles: assess risk across the following key categories—ICT, people, equipment, facilities, environment, supply chain, and legal/regulatory compliance
4. Prioritise recovery and mitigation efforts: rank business functions by criticality and vulnerability, enabling informed decision-making for risk reduction and resource allocation.

Risk profiling through BIA

One of the most actionable insights from a BIA is the ability to assign risk profiles across various organisational domains. This defines areas of vulnerability and criticality, creating a clear roadmap for mitigation strategies.

The following table provides a crude H/M/L risk evaluation, but can be more prescriptive as necessary to fit your needs. The objective is to define which dependencies are of concern.

Dependency	Low risk	Medium risk	High risk
ICT	Basic off-the-shelf software Minimal dependency	Minor changes to software / personalisation Moderate dependency on service providers Moderate use of cloud-based services Legacy systems Stable environment with minimal change	Complex environment Various systems High dependency on managed service providers (MSP)/vendors Bespoke / adapted significantly for the organisation Volatile environment (significant changes occur) Significant use and dependency on cloud-based solutions
People	Large resource pool Easy to resource	May take time to hire the required skill sets	For SMEs Heavy reliance on very specific skill sets Very difficult to replace
Equipment	Standard equipment Easy to procure	Use of specialist equipment, but easy to procure / inexpensive	Equipment expensive / difficult to procure or maybe of limited production
Facilities	Can operate from any location	High dependency on facilities but can be replicated/replaced easily	High location dependency Expensive or not possible to duplicate/replicate elsewhere in a short time
Environment	Standard office environment Few to no hazards	Moderate OH&S/hazards in local environment	Use/storage of chemicals Significant OH&S / hazards / local environment
Supply chain	Standard services Minimal engagement with suppliers/providers of goods and services	Some key business activities supported by service providers	Highly dependent on supply chain and providers of goods and services
Legal and regulatory	Standard legal requirements	Above standard requirements but subject to self-monitoring Minimal reporting requirements	Activities highly regulated and audited for compliance Regular reporting and collection of evidence

Involve key business units, including operations, HR, IT, legal, and supply chain teams, to ensure all critical functions and dependencies are identified, and evaluate the risk profiles for each area, as follows:

Service/activity	Description	Dependency						
		ICT	People	Equipment	Facilities	Environment	Supply chain	Legal & regulatory
Administration	General office administration	Moderate	Low	Low	Low	Low	Low	Moderate
Engineering	Specialist research team	High	High	Moderate	Low	Low	Moderate	High

This activity will structure the conversations around risk management for each business unit, which they are then accountable for tracking and managing. For example, engineering should have the following risks regularly addressed and reported on within its risk register:

- Use of ICT
- People
- Equipment
- Supply chain
- Legal and regulatory

Organisational vulnerabilities

The BIA identifies organisational vulnerabilities related to dependency loss. The table below lists examples of dependencies, issues faced, and potential risk descriptions.

Dependency	Key issues	Risk descriptor
Facilities/equipment	Extensive large equipment used to create products Specialist/costly equipment Facilities across 3 locations/countries	Difficult or costly to replace Different regulations/standards (depending on geography), adding complexity to business processes
People	High dependency on SMEs with specialist training Training takes at least 3 years to perfect Remote workers & WFH High-value procurements	Key-person risk Loss of knowledge Difficult to replace specialists Loss of equipment Insider threat (fraud, espionage, user error)
Information	Process-sensitive personally identifiable information (PII) Sharing PII Creating data/IP	Different regulations/standards (depending on geography and type of service) Loss of data confidentiality, integrity, or availability Data breach (privacy) Loss of IP / competitive advantage
ICT	Significant legacy equipment	Inability to patch or keep maintained Difficult to replace or procure
Supply chain	Reliance on specialist supplier for widgets	Loss of supplier would impact operations

By systematically assessing organisational dependencies and risk factors, such as data security, legacy systems, and supply chain vulnerabilities, businesses can sharpen their understanding of potential disruptions. This approach helps leaders identify where strategic objectives could be jeopardised and informs proactive planning to mitigate risks and strengthen organisational resilience.

> "Risk management is less about avoiding risk
> and more about understanding it."
>
> Unknown

TACTICAL AND OPERATIONAL CTI

When developing a comprehensive view of your organisation's cyber risk profile, you need to leverage intelligence about the organisation from existing cybersecurity activities. Investigate how the technical team currently collates information regarding emerging threats and vulnerabilities to your information systems. Ideally, they will at least be collecting data from specialist interest groups or vendor news to identify and prioritise system vulnerabilities.

But this is only a fraction of what can be done to increase cyber resilience. CTI gained its own domain in ISO 27001:2022, which is a significant recognition of its importance to cybersecurity risk management. CTI is structured by 3 pillars, covering strategic, tactical, and operational intelligence.

Strategic CTI incorporates what we've already learned from the previous chapters: it sets the context of the organisation and indicates what threats are most likely in your organisation based on industry, operations, goals, and what attackers may gain from targeting you. The other 2 pillars are described below.

Tactical CTI

Tactical CTI is an approach that emphasises the need to evaluate the TTPs employed by threat actors. It enables organisations to develop precise remediation and defensive strategies at the technical level which are tailored to counteract these threats effectively.

For example, tactical CTI might uncover an increase in attacks where threat actors spoof HR emails to target payroll systems. This intelligence highlights the need for bolstered email-filtering systems and enhanced employee training to recognise and report phishing attempts.

Effective tactical CTI relies on data collection from a wide range of sources, ensuring that insights are comprehensive and actionable. These sources include:

- Open-source threat intelligence (OSINT): publicly available information on potential threats
- Dark web monitoring: insights into threat actor activities and potential vulnerabilities being discussed or sold
- Industry-specific reports: tailored intelligence focused on threats relevant to your sector.

External attack surface management

EASM plays a critical role in tactical CTI by detecting and mitigating risks associated with your organisation's external digital assets. It involves identifying, analysing, prioritising, and remediating vulnerabilities exposed in your online infrastructure.

EASM gathers information from publicly exposed assets and connections, providing a clear view of how your organisation appears to potential attackers. This includes monitoring for vulnerabilities, misconfigurations, and other risks using OSINT sources such as:

- Websites and blogs
- Social media platforms
- Security news and aggregator websites
- Vulnerability databases (e.g., National Vulnerability Database [NVD], Common Vulnerabilities and Exposures [CVE])
- Dark web forums and marketplaces
- Bug bounty platforms
- Academic research and industry reports
- Cybersecurity podcasts and webinars.

One practical application of tactical CTI is evaluating what information about your organisation and employees is accessible online. This process provides insights into:

- Social-engineering vulnerabilities: how susceptible your employees are to phishing and other social-engineering attacks
- Cybersecurity awareness: the effectiveness of your

organisation's security awareness training and phishing simulation exercises.

For instance, if public data reveals detailed employee profiles or operational processes, this might indicate a need to strengthen awareness training, limit information shared on social platforms, or restrict publicly available details about your systems and infrastructure.

Tactical CTI is a cornerstone activity for understanding an organisation's cyber risk profile because it offers actionable insights into the specific threats that the organisation is most likely to face. By identifying and analysing the TTPs used by threat actors, tactical CTI helps organisations examine their vulnerabilities, evaluate their current exposure, and identify potential attack vectors. This intelligence directly informs the risk profile by revealing how external threats align with internal weaknesses such as misconfigured systems, outdated software, or gaps in employee awareness.

Moreover, by leveraging tools like EASM and OSINT, tactical CTI provides a clear picture of how the organisation is perceived by attackers. This includes identifying publicly available data that may be exploited, such as employee details or system misconfigurations, and assessing the organisation's susceptibility to targeted attacks like phishing or social engineering. These insights allow organisations to prioritise defences where they are most needed, reducing overall risk and ensuring resources are allocated efficiently.

Understanding these specific threats and exposures enables the organisation to adapt its strategies dynamically, aligning with its broader risk-management objectives and improving resilience.

Operational CTI

Operational CTI is where the actionable insights derived from strategic and tactical CTI converge, driving practical measures to fortify the organisation's cyber defences. This stage focuses on operationalising the findings by integrating them into daily workflows, technical configurations, and response mechanisms to address the identified risks proactively.

This is not just about recognising the most relevant threats, but also about enabling the organisation to respond effectively in real time to emerging threats.

By mapping the attacker methodologies, tools, and technologies to the organisation's current environment, a business can assess its risk profile from a highly technical and immediate perspective. This includes identifying how threat actor tactics align with known vulnerabilities, determining the probability of exploitation, and defining the potential impact on critical business processes.

Key activities at this stage include:

1. Evaluating the relevance and likelihood of threat occurrence

   - Leverage the insights from tactical CTI to determine which threats are most likely to target the organisation based on its sector, geographical region, and operational footprint.
   - Develop a risk matrix that categorises threats by likelihood and impact, ensuring that mitigation strategies are focused on the highest priority risks.

2. Prioritising cybersecurity improvements

   - Use industry-specific threat intelligence, such as IOCs or threat reports, to identify and close gaps in the environment.
   - Tailor investments in security tools and practices to address the most pressing risks, ensuring that every improvement is both targeted and effective.

3. Deploying and configuring technical tools

   - Operational CTI enables organisations to select and configure technologies such as SIEM solutions and endpoint detection and response [EDR] systems. These tools serve as the frontline of defence, improving visibility, detection, and response to known threats.

- Establish real-time monitoring capabilities that can ingest intelligence feeds, correlate data, and trigger alerts aligned with the organisation's risk profile.

4. Customising cybersecurity incident-response plans

- Operational CTI confirms that incident-response plans and playbooks are not generic, but instead reflect the unique threat landscape of the organisation.
- Regularly update these plans based on the latest intelligence, ensuring alignment with evolving risks and stakeholder expectations.
- Test and refine these plans through tabletop exercises or simulated incidents to maintain readiness.

The insights derived from operational CTI serve as a critical input to the organisation's risk profile by continually refining its vulnerabilities and threat landscape.

For instance:

- Operational CTI provides a near-real-time view of risks, ensuring that the organisation's risk profile remains relevant in the face of changing threats.
- Insights gained from monitoring tools and incident-response activities can feed back into strategic and tactical CTI, improving the overall accuracy and relevance of the organisation's threat-intelligence program.

Role of threat-hunting programs in risk profiling

Threat hunting is a proactive process that involves identifying and mitigating potential threats within an organisation's environment before they cause harm. By linking the outcomes of tactical CTI, threat hunting becomes more targeted and effective, leveraging insights about attacker TTPs.

Threat-hunting programs offer another opportunity for the organisation to examine its risk profile over time by identifying potential vulnerabilities and weaknesses before they are exploited.

This activity helps to validate the findings from tactical CTI and uncovers unknown risks within the environment such as unpatched vulnerabilities, misconfigurations, or unnoticed anomalous behaviours. These findings allow the organisation to proactively reassess its risk levels for specific assets or processes that may be targets and strengthening its resilience to a cyber attack.

Insights from threat hunting can highlight gaps in detection tools, processes, or policies, leading to improvements in resource allocation, investment in specific security technologies, and better alignment with the evolving threat landscape.

By continuously aligning threat hunting with CTI insights, organisations can maintain dynamic, accurate, and actionable risk profiles.

CTI metrics and reporting

Enhance your ROI and transform knowledge into action by setting clear CTI measures and metrics.

Having made your way through the chapters so far, hopefully you now have more clarity on where risk profiling fits into your cybersecurity program and how the different aspects of profiling, including CTI, refine your cybersecurity strategy.

The metrics are likely to change over time as you learn more about what you want to know and how it can be measured. But where do you start?

As quoted by Stephen R. Covey, you must: "Start with the end in mind!"

What is the outcome you seek and how will you know you've succeeded? This leads us nicely into OKRs...

CTI objectives and key results

For those interested in OKRs, *Measure What Matters* by John Doerr and *Security Metrics: Replacing Fear, Uncertainty, and Doubt* by Andrew Jaquith are highly recommended as complementary resources to this chapter.

A very clear outcome of CTI activities is that we want to build cyber resilience capabilities. To do this we need to enhance our ability to anticipate, prepare for, and respond to changes in our environment (internal and external) so that we can react in a timely manner and protect the organisation from known threats and vulnerabilities.

Our guiding objective is therefore to build cyber resilience. We can achieve this by:

- Strengthening collaboration with key stakeholders
- Enhancing threat visibility across the organisation
- Improving the quality and actionability of CTI
- Optimising threat hunting and analysis capabilities
- Measuring and demonstrating the value of CTI programs
- Enhancing proactive threat defence
- Improving threat intelligence automation
- Enhancing strategic intelligence for long-term risk management.

Objectives need to be *measurable (SMART)*, so we can demonstrate how they will be achieved. This is facilitated by the key results (KRs) that we select based on timeframes, available resources, and overall maturity of the organisation.

The table in Appendix F provides some sample OKRs that should be adapted for your needs and capabilities.

CTI continual improvement

The threat environment is constantly evolving, requiring agility in your approach to CTI. Each emerging threat, new technology, and lesson learned from past incidents offers an opportunity to refine and improve your processes. CTI must not remain static; it thrives on adaptation and commitment to continuous improvement.

Regularly evaluate the effectiveness of your CTI activities to ensure you are gathering the right intelligence at the right time and, most importantly, deriving actionable value from it. The true power of CTI lies not in the information collected, but in how that intelligence is applied to drive decisions and improve resilience.

Embed the insights and outcomes of CTI into your broader cybersecurity strategy. This fosters assurance that the strategy remains relevant, responsive, and aligned with the organisation's needs. A proactive and adaptive cybersecurity strategy demonstrates a clear understanding of both current and emerging risks.

For organisations with a certified ISMS, the annual management review provides a structured opportunity to assess significant internal and external changes. This process evaluates how such changes impact the organisation's cybersecurity objectives and informs necessary adjustments to the ISMS and its supporting strategies. By aligning CTI outcomes with this review, organisations can confirm their cybersecurity approach remains not only compliant, but also strategically effective.

Make continuous improvement a foundational element of your CTI program. Use feedback from monitoring, incident response, and strategic reviews to refine threat intelligence practices and better anticipate risks. This iterative approach helps keep defences ahead of emerging threats and strengthens the organisation's overall risk posture.

CONTROL MATURITY ASSESSMENT

Immature cybersecurity controls skew an organisation's risk profile, making it difficult to accurately identify, prioritise, and address risks. If an organisation is immature in its approach to cybersecurity, it is less likely to fully grasp or measure its vulnerabilities, leading to underestimation of threats, weaknesses, and potential impacts.

Without robust mechanisms in place, the following issues may emerge:

- Attack surfaces are broader, increasing the likelihood of breaches.
- Vulnerabilities remain undetected, creating blind spots in the risk profile.
- Threat actors may exploit these weaknesses, leading to cascading consequences across systems.

Why control maturity assessment is essential

A must-do for any organisation is a current state assessment of its cybersecurity controls to establish a baseline. This is not a one-off exercise—it must be a continuous process integrated into the organisation's broader risk-management approach.

Conducting a control maturity assessment provides several key benefits:

Identifying gaps	It highlights deficiencies in existing controls, ensuring risks are neither underestimated nor overlooked.
Supporting risk-based decision-making	By mapping the maturity of controls against known threats, organisations can align investments with the most critical security gaps.
Regulatory and compliance alignment	Many industry regulations require demonstrable control effectiveness; a maturity assessment helps identify compliance gaps.
Enhancing incident response readiness	Understanding control weaknesses allows for more effective response planning, reducing downtime and impact.

Integrating control assessments with risk profiling

A control maturity assessment should not be conducted in isolation. Instead, it must form part of a comprehensive risk-profiling process. If an assessment is performed without considering the broader risk environment, prioritisation and ROI evaluations will be significantly hindered.

To integrate control assessments effectively, consider the following:

Conduct a gap analysis	Compare the current state of controls against industry best practices (e.g., NIST, ISO 27001, CIS Controls).
Prioritise findings based on risk impact	Not all gaps pose equal risk. Assess which weaknesses expose the organisation to the most significant threats.
Develop a roadmap for control maturity improvement	Set realistic goals for enhancing controls over time, considering resource availability and business objectives.
Align with business and IT strategies	Cybersecurity investments should support broader organisational goals to ensure buy-in from executives and stakeholders.

Common pitfalls in control maturity assessments

When evaluating security controls, organisations often make common mistakes that can lead to gaps in risk management and ineffective security strategies. Here are some key pitfalls to avoid:

1. Do not treat security control assessments as a one-time activity

Conducting assessments only during audits or compliance checks, rather than as part of a continuous security improvement process, leads to a check-box type exercise. The issue with this limited approach is that the focus is purely on "getting through the audit" rather than using the information to direct or prioritise the improvement plan.

The assessments should be utilised as a benchmarking exercise and revisited regularly (timescales depending on the rate of change) to demonstrate progress against a plan and ROI.

- Define the benchmark (initial assessment).
- Agree on the desired target state and create an improvement plan.
- Set the OKRs for the program of work and periodically check in against the plan.
- Regularly review the scope and coverage of controls—

influenced by changes to regulations, standards, stakeholder needs and expectations, and evolving threats.

2. Do not focus only on compliance, rather than security effectiveness

Avoid prioritising compliance checkboxes and relying on policy documentation rather than actual performance testing of controls, security maturity, and resilience. This is a common mistake I've seen often in my role as consultant over the years, where a gap analysis is completed against a specific control framework (say ISO 27001) with the focus purely on the controls within Annex A, without considering:

- Relevancy: are the controls even applicable to the organisation?
- Completeness: has the organisation considered all relevant controls, beyond the Annex A controls, such as PCI DSS, CPS 234, or regulatory requirements?
- Depth: where should the organisation conduct a deep dive into the most important (key) controls?
- Scope: how are the controls applied throughout the organisation? Many assessments are done at the highest level, considering the design elements of a control but not delving into control effectiveness.

Although compliance is an important consideration, the organisation should obtain assurance that the controls are not only designed appropriately, but also operate effectively and have the desired result.

The effectiveness of controls can be stress-tested through various means such as conducting red-teaming events utilising real-world attack scenarios (established through tactical CTI), threat modelling, and data-breach simulations.

Utilise KPIs and metrics to track the performance of security controls.

3. Do not ignore context and business relevance

Avoid applying a generic, one-size-fits-all approach without considering the organisation's unique risk profile. I've seen this time and time again, hence the inspiration for this book! Context cannot be ignored. Without refining what the organisation's unique needs are, it will be impossible to embed a culture of information security as that would not be meaningful and would remain an "IT issue".

4. Avoid underestimating third-party and supply chain risks

Many businesses, regardless of size, have historically assumed that vendors and third-party providers have adequate security in place, without proper validation. We've seen many data breach events caused by third-party and supply chain risks. Businesses have finally (in recent years) recognised their exposure and are now putting greater emphasis on managing their third parties.

Third-party risk management is still immature in practice, as we see businesses depending heavily on certifications (such as ISO 27001, SOC 2 Type II) to satisfy their requirements, when they should delve deeper into the control requirements for their own use case. That is based on the types of products and services provided, how they are used, and their value to the organisation concerned.

When assessing third-party controls, avoid generalisation. ISO 27001 certification is not a clearcut "pass mark"—although it is a good litmus test of a third party's approach to information-security governance, it does not indicate that the controls are appropriately applied for your unique use case.

Refer to Chapter 6 for details on conducting third-party risk assessment.

5. Avoid overlooking asset inventory and visibility gaps

Failing to have an accurate and up-to-date inventory of assets can lead to unprotected systems. It is the old saying: "You can't protect what you can't see"! It's essential to explore your entire environment.

Enhance knowledge of the environment by maintaining a real-time asset inventory and confirm that all critical assets, including shadow IT and third-party integrations, are accounted for.

ENHANCING VISIBILITY

Just as with other aspects of risk profiling, organisations must fully appreciate the scope of what they are protecting. Achieving this requires confidence in their ability to gain visibility over the technical environment, understand interdependencies, track data flows, and be alerted to any environmental changes.

Four steps to improve visibility and define your technical risk profile

1. Define system topology: create a detailed map of systems, databases, firewalls, and storage locations. For each asset, document:
   - IP addresses and physical locations
   - Internet-facing exposure
   - Software versions and configurations
   - Dependencies, including third-party vendors.
2. Map asset integrations: identify how organisational assets interact with external systems to examine where data is stored and processed beyond internal controls.
3. Define control boundaries: clearly outline what the organisation can and cannot control. This helps in assessing any reliance on external parties for security.
4. Validate assumptions through testing: conduct vulnerability scans and penetration tests, particularly on internet-facing assets, to ensure visibility gaps are identified and mitigated.

This information not only helps in defining what needs to be protected, but also provides a clearer picture of where the environment is most vulnerable. Mapping the organisation's external attack surface offers critical insights into which vulnerabilities and exposures require immediate action.

Benefits of using a tool to map network and information flows

Using a tool to map networks and information flows provides several advantages that enhance cybersecurity, risk management, and operational efficiency.

Enhanced visibility & situational awareness	This provides a clear and accurate view of an organisation's infrastructure, including devices, servers, databases, and data flow paths.
	It helps identify unauthorised, unknown, or shadow IT devices that could introduce security risks.
Improved risk profiling & security posture assessment	This enables better identification of critical assets, data repositories, and potential weak points.
	It assists in prioritising security investments by pinpointing high-risk areas.
Faster incident response & threat detection	This helps security teams quickly trace malicious activity and determine how attackers could move within the network.
	It speeds up the containment and remediation process during cybersecurity incidents.
Strengthened compliance & auditing	This provides evidence of security controls and data-protection measures for compliance audits (e.g., ISO 27001, NIST, GDPR, KSA PDPL).
	It validates proper data-flow documentation to comply with regulatory requirements.
Optimised network performance & efficiency	This identifies bottlenecks and inefficiencies in network infrastructure and application performance.
	It improves capacity planning and helps avoid unnecessary hardware/software costs.
Better third-party & supply chain risk management	This maps interdependencies with external vendors, cloud services, and third-party integrations.
	It identifies risks associated with data transfers, APIs, and remote connections.
Informed decision-making for IT & security strategy	This supports scenario planning and impact assessment for cyber threats, system outages, or IT changes.
	It helps executives and security leaders make data-driven investment decisions.

SUMMARY: INTEGRATING THE COMPONENTS INTO RISK PROFILING

Finally, bring all this knowledge together to create a comprehensive and meaningful risk profile for your organisation.

1. Combine findings: use insights from PESTEL, BIA, CTI, and maturity assessments to map the full risk landscape.

2. Quantify risks: assign likelihood and impact ratings to identified vulnerabilities and threats, not forgetting to add known metrics and research statistics to back up your likelihood and impact ratings.
3. Communicate results: share findings with stakeholders to garner alignment and support for mitigation efforts.

"If the initial analysis of risk is not based on meaningful measures, the risk-mitigation methods are bound to address the wrong problems."

Douglas W. Hubbard

CHAPTER 13
RISK APPETITE AND TOLERANCE

Given the information obtained from risk profiling, the organisation should review its risk-appetite statements and define its tolerance levels. This too should be a regular activity, at least once every two years or when significant organisational change occurs.

Cyber risk appetite is an expression of the type and amount of risk the organisation is prepared to take. This promotes consistent, risk-informed decision-making aligned with strategic aims and it also supports robust corporate governance by setting clear risk-taking boundaries.

An example high-level risk-appetite statement:

"Elev8 Resilience is committed to delivering secure, efficient, and innovative technology and cybersecurity solutions.

While we prioritise operational continuity, regulatory compliance, and reputational integrity, we recognise the need to embrace moderate to high levels of risk in technology adoption and innovation to achieve our strategic objectives.

We aim to balance these risks carefully with the imperative to maintain trust, safeguard individual wellbeing, and ensure the resilience of our key products and services."

DEFINITIONS

Term	Definition
Risk appetite	This refers to as the amount and type of risk that an organisation is willing to accept to achieve its objectives in each risk category. It is a high-level strategic decision that reflects the organisation's overall attitude towards risk. The risk appetite is supported in practice through policies and procedures.
Risk tolerance	This is defined as the acceptable amount of deviation from the level set by the risk appetite. It is typically communicated in quantitative terms to set the boundaries of risk-taking. It is also typically closely tied with a specific risk category or activity. Anything outside of risk tolerance will need to be acted upon according to the cyber risk framework.
Risk category	This refers to the type of cyber risk: people, reputational, operational, financial, technology, assets, cloud services, compliance, third party.

There are a few mechanisms available to determine the acceptable levels of risk for your organisation. The most important is to acknowledge your key stakeholders and "interested parties" (as defined by ISO 27001) and what they expect from your organisation in terms of information security, cybersecurity, and privacy. If you've followed the structure of this book, you will already have at least determined who the stakeholders are from Chapter 8.

Create a stakeholder profile in terms of risk appetite: what would they be willing to support and what actions would go against what they believe in? Consider the culture of the organisation, what and who the stakeholders are invested in (product, service, image) and what intangibles are important.

For example, stakeholders may expect your organisation to be innovative and are happy with the exploration of new products and services; this is the benefit of "positive" risk evaluation (what we can do). But it comes with conditions: what are the boundaries of that innovation?

For example:

Positive risk	Constraint
Support development of new products so the organisation stays at the forefront of innovation.	Products must apply security by design. There is zero tolerance for exposing client data.

USING RISK RATINGS TO DEFINE RISK APPETITE

Knowing our risk appetite helps us to define how we manage the identified risks. For example, the following risk-matrix table clearly defines what level of risk can be accepted, what can be tolerated, and what are unacceptable and require immediate attention.

Level of risk	Risk evaluation	Description
Low	Acceptable	Risk can be accepted without further action.
Medium	Tolerable under control	Risk may be accepted but should be monitored regularly and continual improvement sought over the medium and long term: 6–12 months.
High	Unacceptable	Measures for reducing risk should be taken in the short term: 1–3 months.
Extreme	Unacceptable	Immediate action required: 1–4 weeks.

If you do not already have definitions for your risk ratings, in terms of when action must be taken and when, adopt the above example to challenge your existing risk matrix. Do the results add up?

Does your risk matrix support your risk-appetite statements?

Reassessing the risk matrix to align with risk appetite

As your organisation evolves, so too should your risk matrix. It's important to regularly reassess whether the matrix still reflects your current risk appetite, particularly in relation to cybersecurity and whether it is achieving the desired outcomes. For instance, if your organisation is more risk averse—meaning it has low tolerance for cyber threats—your matrix may feature a higher number of amber and red ratings (representing high and extreme risks). This would signal a need for more immediate or intensive risk-mitigation strategies.

The risk matrix should never be treated as static. Static matrices can inadvertently undervalue significant risks, especially to critical assets, or conversely inflate risk ratings to the point where everything seems urgent, making prioritisation difficult and diluting your ability to act strategically. Be careful not to fall into the trap of overrating all risks as high or extreme. While this might appear conservative, it can limit business agility, lead to over monitoring, and divert attention away from the risks that truly matter. Striking the right balance is essential to supporting informed, timely decisions without stifling innovation or exhausting resources.

The strength of a risk matrix lies in its flexibility—so long as its criteria are consistently defined. It's entirely appropriate and often necessary to adapt the matrix for specific assets, environments, or scenarios, particularly when dealing with high-value or high-impact assets. These may include systems that are expensive to build or maintain, that store sensitive business information, or that hold personal or health data.

	Rare	Unlikely	Possible	Likely	Almost Certain
Critical					
Major					
Moderate					
Minor					
Insignificant					

For example, as shown above, a risk-averse organisation with critical information assets might adopt a matrix where even a possible likelihood results in elevated monitoring, especially if the impact would be more than minor. This would confirm that key assets are continually reviewed and quickly defended when new threats arise.

	Rare	Unlikely	Possible	Likely	Almost Certain
Critical					
Major					
Moderate					
Minor					
Insignificant					

By contrast, as shown above, an organisation with a higher risk appetite or assets of lower strategic importance might use a matrix that reflects greater tolerance. In this case, moderate risks (those that are only possible and have limited impact) might not require intensive oversight. Monitoring and mitigation would only escalate as likelihood and impact increased.

Whether adapting your matrix to different departments, asset classes, or strategic objectives, always verify that your definitions of high and extreme risk remain consistent. These thresholds dictate not just how risks are ranked, but how quickly your team must act—and misalignment here can lead to confusion or missed opportunities for timely intervention.

"By failing to prepare, you are preparing to fail."

Benjamin Franklin

CHAPTER 14

USING RISK PROFILING TO ENHANCE INCIDENT RESPONSE

An effective risk profile informs incident-response planning as it establishes which incidents would matter most to your organisation and how they might eventuate. Not every incident is created equal and risk profiling lets you triage what needs urgent attention.

A well-articulated risk profile allows faster decision-making because key scenarios are already understood and mitigation strategies have been discussed in advance.

Incident response is where all your preparation either pays off or falls short. A strong, actionable risk profile ensures your organisation doesn't waste time during a crisis trying to figure out what matters—it already knows. This aligns people, processes, and plans ahead of time so that decisions can be made quickly, confidently, and in the best interests of the organisation.

When the dust settles, your risk profile should be the first document you revisit, not just to update it but to embed the lessons learned in how you manage future threats.

The following are critical inputs that enable the incident-management processes to stay up to date and relevant for your organisation:

Tactical and operational CTI	By utilising methodologies such as CTI where we investigate the likely TTPs threat actors may use, we can use that intelligence to develop risk-informed playbooks and containment strategies for the most likely or highest impact risks identified in the organisation's profile.
Historical data	We can utilise past incidents or events from other organisations to learn not only how an event occurred, but how the organisation responded. This should be used to test assumptions within our plans and to review and update risk-cause events and impact ratings so incidents are properly categorised.
Risk indicators	Risk profiling defines how key risks could materialise (e.g., through credential compromise, vendor failure, insider threat) and should directly shape the response scenarios you prepare for.
Vulnerabilities	Profiling reveals the controls you're relying on. If an incident involves a known weak point, response plans should account for likely escalation paths and alternative measures.
Testing	Risk profiles should be used to develop real-world scenarios for simulation exercises, which in turn can test and validate assumptions within the profile itself.
Lessons learned	Close the feedback loop! After each incident, review the risk profile. Was the incident something you had identified? If not, why? If yes, was it given appropriate priority? Post-incident reviews often expose failed controls or untested assumptions. Use this information to update both your risk register and your control maturity assessments.

BUILDING RESILIENCE BY ALIGNING RISK PROFILING WITH BUSINESS CONTINUITY PLANNING

Often, incident response is siloed from business continuity. A good risk profile bridges that gap, connecting cyber threats with operational impacts (e.g., to supply chains, safety, revenue). By clearly defining the organisation's risks, this information can be used to develop realistic, organisation-specific cyber crisis scenarios that are reflected and managed through both incident response and business continuity plans.

By defining stakeholder needs and expectations early in the development of plans, we can uphold that we are not only managing risks but making sure the response and recovery plans enable us to meet their expectations for recovery and inform the required communication channels.

The results of a comprehensive BIA not only help identify the most critical information assets, business processes, and dependencies, but offer an opportunity to evaluate how these may be vulnerable and how they may fail during a cyber event.

"Everybody has a plan until they get punched in the mouth."

Mike Tyson

CHAPTER 15
THE BOARD'S ROLE

One of the essential functions of a board is to set the organisation's risk appetite and confirm that the risks it faces are clearly identified and managed. The board plays a pivotal role in supporting the risk-management process by continuously monitoring risks and challenging the organisation to adapt its approach to an evolving threat landscape. By asking the right questions, board members validate that current risks are effectively managed and potential future risks are anticipated, particularly in rapidly advancing sectors like technology and cybersecurity.

This responsibility becomes even more critical as we witness exponential growth in AI, advancements in technology, increasingly persistent cyber threats, and evolving regulations across the globe. As technology reshapes industries, boards must confirm that risk-management processes align with these developments by regularly reviewing and updating the organisation's risk profile.

RISK PROFILING AND CYBERSECURITY

Continuous risk profiling should be a core part of strategic decision-making. It's essential for the board to acknowledge the organisation's overall cyber risk profile so it can make sure that the risks are being addressed, remain relevant, and align with broader business objectives.

In a dynamic threat landscape, a static risk profile can quickly become outdated. Board members must verify that the organisation has the systems in place to continuously monitor, update, and adjust its risk profile using real-time data and threat intelligence.

The board's role in cyber risk management includes:

- Proactive oversight that encourages a culture of cyber risk awareness
- Support of strategic investment in cyber resilience to protect the organisation
- Consideration of the long-term vision for cyber resilience, in line with the organisation's strategic objectives and future growth plans.

By focusing on continuous risk profiling, the board can provide essential oversight and help management proactively address both current and future threats.

Proactive oversight

Board members can promote a culture of proactive risk management by fostering communication between IT, security, and business units. In a world of emerging technologies, it's critical for the board to comprehend both short-term and long-term risks. For instance, while AI may increase operational efficiency, it can also introduce risks such as AI-driven attacks. Similarly, quantum computing threatens to undermine encryption protocols, which must be addressed in long-term planning.

Many emerging technologies rely on third-party providers (e.g., cloud services, IoT manufacturers), increasing an organisation's exposure to supply chain risks. The board must confirm that third-party risks are continuously integrated into the overall risk profile and managed with the same rigour as internal risks.

Fostering collaboration across organisations and disciplines is essential; by working together, everyone plays an active role in identifying, communicating, and addressing threats and vulnerabilities. This collective approach strengthens proactive oversight and empowers all involved to contribute to a more resilient security posture.

Regulations and standards

Staying ahead of evolving regulations, such as privacy laws and cybersecurity frameworks, is essential for avoiding fines and reputational damage. Boards must ensure that risk profiles are updated to include regulatory risks, particularly as new technologies introduce compliance challenges.

Strategic investments in resilience

The board's primary responsibility is to set the tone from the top, instilling a culture where resilience and proactive risk management are foundational values. By modelling this commitment, board members shape behaviours throughout the organisation, encouraging vigilance and adaptability in the face of emerging threats.

To build such a culture, the board should champion strategic investment in advanced security solutions, such as AI-powered threat detection, to enable continuous risk monitoring. It should also prioritise ongoing employee training, recognising that human error often represents the greatest vulnerability in cybersecurity.

Continuous risk profiling requires sustained investment in technologies that automate vulnerability assessments, real-time monitoring, and threat detection. Through active oversight and by fostering the right organisational mindset, the board verifies that resilience is woven into the fabric of the business, enabling the organisation to safeguard its long-term future.

Long-term vision for cyber resilience

The evolving landscape of technology, cybersecurity threats, and regulatory requirements significantly heightens the board's responsibilities.

Boards must recognise that effective risk management is no longer a one-time exercise, but a continuous process.

Board members should challenge management to look beyond immediate threats and build a forward-looking risk profile that considers emerging risks such as quantum computing and AI, ensuring that the cyber risk profile aligns with the broader business objectives, particularly in relation to digital transformation—this is critical to long-term resilience.

It's the role of the board to:

- Actively oversee the development and adaptation of risk-management frameworks, ensuring risks are regularly identified, monitored, and addressed with agility
- Make sure risk profiles are updated dynamically, incorporating real-time data and anticipating both internal and external threats, including those posed by emerging technologies and third-party providers
- Ensure the organisation's digital strategy incorporates cybersecurity and privacy considerations e.g., if the business is expanding into IoT or AI, there must be specific risk profiles for these technologies, as they introduce new attack vectors
- Investments in advanced cyber resilience solutions and ongoing employee training are prioritised, recognising that both technology and human factors contribute to organisational vulnerability
- Stay ahead of evolving legal and compliance obligations by ensuring risk profiles reflect regulatory changes and potential compliance risks introduced by new technologies
- Set a tone from the top that promotes a culture of vigilance, collaboration, and adaptability, making resilience and proactive risk management foundational values throughout the organisation.

By embracing these principles, the board not only safeguards the organisation's assets and reputation, but also empowers it to harness new technological opportunities and confidently navigate future challenges. In an era of rapid digital and regulatory change, active board engagement in agile, forward-thinking risk management is essential. Through vigilant oversight and continuous adaptation, board members can steer their organisation towards resilience, security, and sustained strategic growth.

"The essence of strategy is choosing what not to do."
Michael Porter, economist, author, and professor

CHAPTER 16
CONTINUAL IMPROVEMENT

Continual improvement sits at the core of effective risk management, evolving the process from a static checklist into a living cycle of adaptation and resilience. Rather than relying solely on periodic reviews, organisations must embed continual improvement by ensuring their practices and controls shift in step with new technologies, changing business objectives, and regulatory developments.

A forward-thinking approach integrates real-time threat intelligence into daily operations. This allows risk profiles and controls to be adjusted promptly as the threat landscape shifts, maintaining ongoing relevance and robustness. Instead of waiting for scheduled reviews, significant changes in risk exposure can be addressed as they arise, bolstering organisational resilience.

Equally vital is maintaining sharp awareness of the wider business context. Events such as mergers, new market ventures, and major product launches reshape risk profiles and may require fresh mitigation strategies. By sustaining open dialogue with stakeholders and leaders, risk managers reinforce alignment with business priorities and enhance the ability to anticipate change, rather than simply responding to it.

Predictive analytics further strengthens this cycle. Harnessing data and industry insights, organisations can anticipate emerging risks—allocating resources and reinforcing controls before vulnerabilities are exploited. This proactive stance turns risk management into a strategic enabler, rather than a reactive function.

The use of structured methodologies such as PESTEL analysis or BIA brings discipline to the continual-improvement process. Regular reassessment using these tools confirms that all relevant factors—technology, law, economics, and more—are captured in the risk profile and inform decision-making.

Fundamentally, continual improvement is about fostering a culture of curiosity and adaptability. It transforms resilience from a one-time objective into a sustained journey, empowering organisations to seize new opportunities while staying secure amid complexity.

MONITORING RISK AND ADJUSTING THE PROFILE

Organisations are constantly evolving—through growth, changing markets, evolving processes, and new technologies. To remain aligned with current realities, risk profiles must be reviewed and updated regularly. This means that risk prioritisation stays accurate and resources are directed where they are most needed.

Undertaking structured reviews, such as an annual PESTEL analysis, identifies both internal and external factors that might shift the organisation's risk landscape, from economic fluctuations to regulatory changes or geopolitical events.

MANAGEMENT REVIEW AND CONTROL EFFECTIVENESS

An effective ISMS, as defined by ISO/IEC 27001, embeds continual improvement into its core. This includes structured management reviews that assess how well the risk-management framework supports the achievement of strategic objectives.

ISO 27001 highlights this in:

- Clause 9: Performance Evaluation. It requires that "The results of the management review shall include decisions related to continual improvement opportunities and any needs for changes to the information security management system."
- Clause 10: Improvement. Organisations "shall continually improve the suitability, adequacy and effectiveness of the information security management system."

This process is essential for validating whether your risk profiling efforts are working and if they still reflect the organisation's actual risk exposure.

MEASURING WHAT MATTERS—HUBBARD'S CHALLENGE TO RISK MANAGEMENT

If your organisation has never reviewed the effectiveness of its risk-management framework, now is the time to ask:

"What is the evidence for the belief that it works?"

This question, posed by Douglas W. Hubbard in his book *The Failure of Risk Management: Why It's Broken and How to Fix It,* underscores the need to test and challenge the methods we use. It's not enough to have a risk process in place—we must demonstrate that it is meaningfully reducing risk.

Hubbard prompts leaders to think critically:

> "If risk went down after the implementation of a new policy, how would you know? How long would it take to confirm that the outcome was related to the action taken? How would you determine whether the outcome was just due to luck?"

He warns that the most dangerous kind of risk management is not the absence of it, but a method that is used with confidence, never questioned, and might "cause erroneous decisions to be made that would otherwise not have been made."

143

To complement his other work, Hubbard's *How to Measure Anything in Cybersecurity Risk* is a must-read for those seeking quantitative, evidence-based approaches to managing cyber risk.

FROM INSIGHT TO ACTION: METRICS AND KPIS

To move from subjective assessment to objective improvement, organisations should track KPIs and metrics tied to risk-profiling efforts. These include:

- KPIs relating to how well you understand and manage the organisation's risk profile e.g., has the intelligence led to a reduction in cybersecurity incidents, enabled faster incident response, or improved alignment between security and business objectives?
- Measuring the effectiveness of implemented controls, which enables us to track how well current security controls are performing and whether they mitigate the intended risks
- Building feedback loops into your risk-management process to incorporate lessons learned, enabling you to adjust to shifting goals and refine your profile over time.

Ultimately, risk profiling should lead to decisions.

As Hubbard puts it:

> "Relevant risk management should be based on risk assessment that ultimately follows through to explicit recommendations on decisions."

Otherwise, it's just a compliance exercise with no ROI.

"Cybersecurity isn't just IT—it's business resilience."

Stéphane Nappo, Vice President & Global Chief Information Security Officer (CISO) at Groupe SEB

CHAPTER 17
RISK PROFILING FOR SMES

Every organisation, regardless of size, must recognise and address the risks that come with its products and services to satisfy stakeholders and meet industry standards. Yet, not all businesses operate on a level playing field. For small and medium-sized enterprises (SMEs), the challenge is even greater: they must demonstrate sufficient risk management without implementing controls that are prohibitive— such as those that are excessively costly, time-consuming, or technically complex.

For example, a sophisticated enterprise-grade security system might be prohibitively expensive for a small business, both in up-front investment and ongoing maintenance fees. Similarly, policies that require a dedicated in-house cybersecurity team or demand significant staff training hours may simply be unfeasible for a company with limited resources. Overly burdensome compliance requirements or complex multi-factor authentication systems can also stretch an SME's capabilities and budget, ultimately threatening the success or sustainability of the business.

Effective risk profiling may seem daunting but, with the right focus and practical steps, it becomes an achievable and high-impact exercise. Here's an actionable guide.

1. PRIORITISE HIGH-IMPACT RISKS

- Identify your business's most critical risks, such as those affecting finances, cybersecurity, or daily operations.
- Focus energy and resources on the risks that could cause the most harm if left unmanaged.

2. DEFINE RISK APPETITE

- Decide what level of risk is acceptable for your business.
- Use this threshold to shape which risks require immediate attention versus those you can tolerate.

3. USE COST-EFFECTIVE TOOLS & AUTOMATION

- Utilise free or open-source cybersecurity tools (e.g., Snort, OpenVAS, AlienVault) for monitoring and alerts.
- Automate routine security checks, updates, and alerts where possible to save time and reduce errors.
- Leverage existing office software (like M365 or Google Workspace) for risk tracking and communication.

4. KEEP RISK-ASSESSMENT FRAMEWORKS SIMPLE

- Adopt a streamlined and easy-to-apply framework (e.g., a simplified ISO 31000, and basic CVSS scoring).
- Create a minimum viable profile with essential information: key assets, main threats, and biggest vulnerabilities.

5. OUTSOURCE SELECTIVELY AND BUILD NETWORKS

- Outsource specialised tasks (like annual cyber audits or incident-response planning) for expertise without full-time costs.

- Join industry alliances or business groups to access shared resources and threat intelligence.

6. EMPOWER AND TRAIN YOUR TEAM

- Provide all staff with foundational cybersecurity training using free resources.
- Cross-train team members to handle basic risk-management tasks—distribute knowledge to build resilience.

7. PREPARE FOR INCIDENTS

- Draft simple playbooks for common threats (like phishing or data breaches) with clear action steps and responsibilities.
- Establish an escalation chain, knowing who to contact externally in case of a serious incident.

8. MAKE RISK MANAGEMENT PART OF THE ROUTINE

- Conduct brief, quarterly risk reviews to keep your risk profile current and adjust for new threats or business changes.
- Encourage ongoing, incremental improvements, update documentation, add new risks, and refine processes regularly.
- Promote risk awareness in meetings and discussions to embed resilience in your company culture.

9. SHARE AND REPLICATE SUCCESSES

- Document effective controls or responses, apply what works across the business, and use lessons learned to strengthen future practices.

Even with limited resources, smaller businesses can make meaningful progress in risk profiling by focusing on the essentials, leveraging community knowledge, and embedding risk awareness throughout the organisation. By adopting these actionable steps, SMEs lay the foundation for business resilience and a proactive approach to cyber threats.

"The greatest risk in life is not taking one."

Mark Zuckerberg, co-founder, chairman, and CEO of Meta Platforms

CONCLUSION

In today's unpredictable landscape, the ability to anticipate, withstand, and adapt to threats is what differentiates truly resilient businesses. This book has shown that effective risk profiling isn't reserved for large enterprises—it is achievable for any organisation willing to focus on core principles: understanding your risks, acting on what matters, and nurturing a culture where resilience is second nature.

We have explored practical steps, developing straightforward playbooks for incidents like phishing and data breaches, built clear escalation paths, and committed to regular, bite-sized risk reviews. Crucially, resilience becomes embedded not through rigid frameworks, but through iterative improvement, transparent documentation, and open sharing of what works. By making risk management a routine part of your business DNA, you empower your teams to respond to change with agility and confidence.

For me, business resilience isn't just about security—it's about clarity, alignment, and confidence. I love working with teams to uncover their genuine risks and priorities. My role is to help you make informed decisions that underpin your strategic ambitions. When security is woven seamlessly into the fabric of your business, it ceases to be a blocker and instead becomes a catalyst enabling progress, innovation, and unity. That's when an organisation can truly breathe and move forward together.

Some of my clients are seasoned CISOs who need a sounding board or an extra set of hands. Others are technical leaders who've inherited security responsibilities and are seeking the confidence and momentum to lead. I meet organisations where they seek tailored guidance rather than one-size-fits-all frameworks, grounded in practical insight and real-world action.

That's the founding spirit behind Elev8: to blend strategic vision, mentorship, and hands-on delivery so that organisations don't just achieve compliance but cultivate lasting resilience.

By embracing these principles, your organisation can chart a path where risk management is not a burdensome checkbox, but an ongoing, empowering journey. Through community, clarity, and continuous improvement, you set the stage for a safer, stronger future—one where security is not the greatest risk, but the greatest enabler.

As you embark on your journey toward greater resilience, know that you are not alone. Community is the multiplier of progress, and the collective wisdom of like-minded leaders is at your fingertips. I invite you to join our online community at **www.elev8ciso.com**, where practitioners, mentors, and trailblazers share insights, stories, and practical strategies to help one another navigate the evolving landscape of security and resilience. Bring your questions, your challenges, and your experiences—we're building a space where every voice matters and every lesson strengthens us all. Together, let's shape the future of business resilience and turn security into your strongest asset.

APPENDIX A:
GLOSSARY OF TERMS

Advanced persistent threat (APT)

A sophisticated and prolonged cyberattack in which an unauthorised actor, often state-sponsored or well-resourced, gains access to a network and remains undetected for an extended period. APTs aim to steal data, survey, or compromise operations, leveraging advanced techniques to evade traditional security measures and require coordinated, strategic responses from organisations

Artificial intelligence (AI)

A wide range of technologies, from rule-based systems to advanced machine-learning algorithms, increasingly used to support decision-making, automate processes, and enhance user experience across industries

AI impact assessment (AIIA)

An AIIA is a structured process designed to evaluate the potential effects—positive and negative—of implementing artificial intelligence systems within an organisation or for a specific use case. This assessment goes beyond technical feasibility to consider ethical, legal, social, and business implications. It typically involves identifying risks such as data privacy concerns, bias in algorithms, and operational disruptions, while also highlighting opportunities for innovation, efficiency, and competitive advantage.

Application Programming Interface (API)

APIs are a set of rules, protocols, and tools that enables different software applications to communicate with each other. APIs define how requests for data and services are made and how responses are delivered, allowing developers to integrate, extend, or interact with existing systems efficiently and securely. APIs are fundamental for connecting disparate software components, enabling automation, and supporting interoperability across platforms.

Bug Bounty

A bug bounty is a structured, incentive-driven programme where individuals are rewarded for discovering and responsibly disclosing security vulnerabilities. These programmes are typically run by private organisations, government agencies, or through specialised third-party platforms.

Business continuity plan (BCP)

A documented strategy outlining procedures, processes, and resources required to continue essential business operations during and after a disruption or crisis

Business impact analysis (BIA)

A systematic process to identify and evaluate the potential effects of disruptions to critical business operations because of an accident, emergency, or disaster

Business resilience

A company's capacity to anticipate, withstand, adapt, and recover from disruptions, threats, or adverse events, ensuring continuity and progress

Centrally managed database (CMDB)

A CMDB is a centralised repository that stores detailed information about the hardware, software, network devices, people, and other assets—collectively known as configuration items (CIs)—within an organisation's IT environment. The CMDB acts as an authoritative source, capturing relationships and dependencies between these items, which is vital for effective IT service management (ITSM).

Chief Information Security Officer (CISO)

A CISO is a senior executive responsible for developing and over-seeing an organisation's information security strategy. The CISO works closely with other executives and IT teams to promote a culture of security awareness and resilience across the entire organisation.

Continuous improvement

An ongoing process of identifying, evaluating, and enhancing processes, often through iterative steps, to strengthen resilience and security

Core principles

Foundational ideas or values guiding risk-management efforts, such as understanding risks and fostering a resilient culture

Cyber threat intelligence (CTI)

Collection and analysis of information about current and potential cyber threats, enabling organisations to make informed decisions about their security posture and response

CVSS

Data breach

An incident in which sensitive, protected, or confidential data is accessed or disclosed without authorisation

Data loss prevention (DLP)

Data Loss Prevention (DLP) refers to a suite of strategies, tools, and processes designed to detect, prevent, and respond to the unauthorised sharing, transfer, or exposure of sensitive information.

DLP systems typically monitor data in use (on endpoints), in motion (across networks), and at rest (stored on servers or devices). They enforce policies that can block, quarantine, encrypt, or alert when data transfer activities deviate from established security rules.

DDoS

DDoS (Distributed Denial of Service) is a type of cyberattack in which multiple systems, often compromised computers or devices, are used to flood a targeted server, service, or network with an overwhelming volume of traffic. The goal is to disrupt normal operations, making the targeted resources slow, unreliable, or completely unavailable to legitimate users.

Escalation path

A clearly defined sequence of steps or contacts to follow when an incident or risk exceeds the capacity of frontline teams, ensuring appropriate and timely response

External attack surface management (EASM)

External attack surface management (EASM) is the practice of continuously discovering, monitoring, and assessing an organisation's internet-facing assets—such as websites, servers, cloud services, and applications—to identify potential vulnerabilities and exposures that could be exploited by attackers.

Governance, risk, and compliance (GRC)

GRC refers to the integrated collection of capabilities that enable an organisation to reliably achieve objectives (governance), address uncertainty (risk management), and act with integrity (compliance with laws, regulations, and policies). GRC frameworks help coordinate processes across departments to align business strategies, manage risks proactively, and ensure adherence to external and internal requirements.

Indicator of compromise (IOC)

An indicator of compromise (IOC) is a piece of forensic data—such as a file hash, IP address, domain name, or unusual network activity—that suggests a system or network may have been breached or targeted by a cyberattack. IOCs help security teams detect, investigate, and respond to potential threats by providing evidence of malicious activity.

Industrial control system (ICS)

Industrial Control System (ICS) is a general term used to describe the integration of hardware, software, networks, and controls that are designed to monitor and manage industrial processes. These systems are pivotal in industries such as energy, manufacturing, water treatment, transportation, oil and gas, and many more where automation, efficiency, and safety are essential.

Information security management system (ISMS)

An Information Security Management System (ISMS) is a comprehensive framework of policies, processes, and controls designed to systematically manage an organisation's sensitive information, ensuring its confidentiality, integrity, and availability. The ISMS provides a structured approach to identifying, assessing, and mitigating information security risks, allowing organisations to safeguard data against threats such as cyberattacks, data breaches, or inadvertent loss.

Intellectual property (IP)

Intellectual property (IP) refers to creations of the mind—such as inventions, literary and artistic works, designs, symbols, names, and images—used in commerce and protected by law. IP encompasses patents, copyrights, trademarks, and trade secrets, allowing individuals and organisations to control, use, and benefit from their innovations or creative output.

Iterative improvement

A practice of making gradual, repeated enhancements to processes or frameworks, rather than attempting radical change all at once

Key risk indicator (KRI)

A Key Risk Indicator (KRI) is a measurable value used by organisations to signal potential threats to business objectives or operations. KRIs are an essential component of risk management frameworks, enabling decision-makers to monitor trends, anticipate problems, and strengthen their organisation's resilience against various threats.

Large language model (LLM)

A type of AI model trained on vast amounts of text data to understand, generate, and manipulate human language; these models use deep-learning techniques, particularly a neural network architecture called a *transformer*, to perform complex language tasks such as summarisation, translation, answering questions, or writing content

LLM poisoning

An adversarial attack where malicious actors inject harmful or manipulated data into the training datasets used to develop or fine-tune LLMs; the goal is to corrupt the model's behaviour, causing it to produce biased, harmful, or exploitable outputs when specific triggers are encountered

Managed service provider (MSP)

A third-party company that remotely manages a customer's IT infrastructure and end-user systems, often delivering services such as network monitoring, cybersecurity, data backup, and technical support. MSPs enable organisations to outsource routine IT tasks, improve operational efficiency, and access specialised expertise without maintaining a large in-house IT team.

Multi-factor authentication (MFA)

Multi-factor authentication (MFA) is a security mechanism that requires users to provide two or more independent forms of identification before gaining access to a system or account. These factors typically include something the user knows (like a password), something the user has (such as a smartphone or security token), and something the user is (like a fingerprint or facial recognition), significantly reducing the risk of unauthorised access.

Operational technology (OT)

Operational technology (OT) refers to the hardware and software systems that monitor, control, and manage industrial processes, equipment, and physical devices. Unlike information technology (IT), which focuses on the management of data and digital information, OT deals directly with the physical operations of an organisation—often in sectors such as manufacturing, energy, transportation, and utilities.

Phishing

A type of cyber-attack that uses deceptive emails or messages to trick individuals into revealing sensitive information or installing malicious software

Playbook

A documented, step-by-step guide outlining procedures and best practices for responding to specific incidents or risks

Risk management

A systematic process of identifying, assessing, and addressing risks to minimise their impact on an organisation

Risk profile

A comprehensive view of the types, likelihoods, and impacts of risks facing an organisation at a given time

Security culture

The attitudes, beliefs, and practices shared across an organisation that prioritise and support security and resilience

SIEM

SIEM, or Security Information and Event Management, refers to a comprehensive platform that aggregates, analyses, and manages data related to security events across an organisation's IT environment. By collecting logs and information from diverse sources—including servers, network devices, applications, and endpoints—a SIEM system provides centralised visibility into potential threats and vulnerabilities.

Service-level agreement (SLA)

A Service-Level Agreement (SLA) is a formal document or contract between a service provider and a client that outlines the specific standards and performance metrics the provider is expected to meet. SLAs define measurable parameters such as uptime, response times, resolution times, support availability, and the scope of services delivered.

SOAR

SOAR, or Security Orchestration, Automation, and Response, is a set of technologies that helps organisations streamline and enhance their security operations. By integrating different security tools and automating routine tasks, SOAR enables faster detection, analysis, and response to cybersecurity threats. It allows security teams to develop and execute incident response workflows, reduce manual workloads, and improve overall efficiency. Through centralised management and collaboration, SOAR platforms support consistent and scalable security practices, helping organisations respond to incidents with agility and precision.

Supplier, input, process, output, client (SIPOC)

SIPOC is an acronym that stands for Supplier, Input, Process, Output, Client. It is a high-level visual tool commonly used in process improvement and quality management initiatives, such as Lean and Six Sigma. SIPOC provides a structured way to map out and analyse the key elements of a process from start to finish, ensuring all critical components are identified and understood.

VPN

A Virtual Private Network (VPN) is a technology that establishes a secure, encrypted connection over a public or untrusted network—such as the internet—enabling users to send and receive data as if their devices were directly connected to a private network. VPNs are widely used to protect sensitive information, ensure privacy, and allow remote access to organisational resources. By masking a user's IP address and encrypting all transmitted traffic, VPNs help maintain confidentiality, integrity, and authenticity of data while traversing potentially insecure networks.

APPENDIX B:
ADDITIONAL REFERENCES & BEST PRACTICES

NIST Cybersecurity Framework (CSF)

Developed by the US National Institute of Standards and Technology, the NIST Cybersecurity Framework is a widely adopted approach for managing and reducing cybersecurity risk. The framework organises best practices, standards, and guidelines into 5 core functions: identify, protect, detect, respond, and recover. It helps organisations of all sizes to assess their current security posture, establish target states, and communicate risk across stakeholders.

Website: https://www.nist.gov/cyberframework

NIST Special Publication 800-30 – Risk Assessment

NIST SP 800-30 provides comprehensive guidance on conducting risk assessment, offering structured methodologies for identifying threats, vulnerabilities, impacts, and likelihoods. It complements the NIST CSF and is invaluable for organisations implementing formal risk-profiling practices.

Website: https://csrc.nist.gov/publications/detail/sp/800-30/rev-1/final

CIS Controls

The Center for Internet Security (CIS) Controls are a prioritised set of actions and guidelines aimed at mitigating the most prevalent cyber attacks. Organised into implementation groups (IGs) to suit organisations of varying resources and risk profiles, the CIS Controls provide practical steps for improving cyber defence. They are updated regularly in response to evolving threats and are recognised globally as a baseline for effective security.

Website: https://www.cisecurity.org/controls

ISO/IEC 31000: Risk Management Guidelines

This standard provides principles and frameworks for risk management, not limited to cybersecurity. It can be used to align cyber risk management with enterprise risk management, making it ideal for board-level communication and strategic alignment.

Website: https://www.iso.org/iso-31000-risk-management.html

ISO/IEC 27001: Information Security Management

ISO/IEC 27001 is an international standard for establishing, implementing, maintaining, and continually improving an ISMS. It provides a risk-based approach to information security, ensuring confidentiality, integrity, and availability of data. Organisations certified to ISO/IEC 27001 demonstrate compliance with globally recognised best practices and can build trust with clients and partners.

Website: ISO/IEC 27001:2022 - Information security management systems

ISO/IEC 42001: Artificial Intelligence Management System

ISO/IEC 42001 is the first international standard for AIMS, published in 2023. It provides a structured framework for organisations to manage risks and opportunities associated with the development, deployment, and use of AI systems. The standard applies to all types of organisations, regardless of size or sector, that are involved with AI, from AI developers to end-users.

Built on the principles of trustworthiness, transparency, accountability, and human oversight, ISO/IEC 42001 maintains that AI is implemented in a way that aligns with ethical guidelines and regulatory expectations. It incorporates elements of governance, risk management, data integrity, and lifecycle controls, helping organisations build confidence in AI applications internally and externally.

As AI becomes increasingly embedded in cybersecurity tooling and risk modelling (e.g., predictive threat detection, user behaviour analytics, autonomous controls), ISO/IEC 42001 is a valuable companion to existing frameworks like ISO/IEC 27001 and the NIST CSF. It supports the integration of AI within existing risk-management systems, and aligns with broader digital trust and resilience goals.

Website: https://www.iso.org/standard/81230.html

ISO/IEC 27090: AI Guidance for Addressing Security Threats and Failures in AI Systems

ISO/IEC DIS 27090 provides guidance for organisations to address security threats and failures in AI systems throughout their lifecycles. It aims to help organisations understand the consequences of security threats to AI systems and offers descriptions of how to detect and mitigate such threats.

Board-level cybersecurity guidance

Leading organisations such as the World Economic Forum (WEF) and Gartner provide in-depth reports and guidelines to enhance cyber risk governance at the board and executive levels. These resources assist directors and senior leaders in understanding the strategic implications of cybersecurity, fostering a security-focused culture, and ensuring effective oversight of organisational risk.

- WEF publishes influential reports such as "Principles for board governance of cyber risk" and "Cybersecurity leadership principles" aimed at CEOs, CISOs, and board directors.

Website: https://www.weforum.org/agenda/archive/cybersecurity

- Gartner cybersecurity leadership research offers practical insights and strategic recommendations tailored for executives navigating digital transformation and cyber threats.

Website: https://www.gartner.com/en/insights/cybersecurity

FAIR (factor analysis of information risk)

This is a quantitative risk-analysis model that helps organisations prioritise and evaluate cybersecurity risks in financial terms. FAIR provides a structured method for analysing threat scenarios, assessing probability and impact, and enabling data-driven security investments.

Website: https://www.fairinstitute.org

ASD Essential Eight (Australia)

Developed by the Australian Signals Directorate (ASD), the Essential Eight is a set of baseline mitigation strategies designed to prevent malware delivery, limit cyber incidents, and enable fast recovery. Widely adopted in Australian government and corporate sectors, it's often used as a risk-based maturity model.

Website: https://www.cyber.gov.au/acsc/view-all-content/essential-eight

APPENDIX C:
KEY QUESTIONS FOR THE BOARD

In today's rapidly evolving digital landscape, effective oversight of cybersecurity and risk management has become an essential responsibility of any organisation's board. The decisions made at this level profoundly influence how resilient a business will be in the face of emerging threats, technological transformation, and regulatory change. To support informed, strategic governance, board members must regularly challenge assumptions, scrutinise risk profiles, and confirm that investments and policies align with both current and future challenges.

The following key questions are designed to guide the board's discussions, foster a culture of proactive cybersecurity, and empower leadership to make well-informed decisions that protect the organisation's assets, reputation, and long-term growth.

Question

How are we ensuring that our risk profile evolves as new threats emerge and as our business grows?

Do we have systems in place to continuously monitor, update, and adjust our risk profile?

What are the potential cybersecurity risks associated with the technologies we are adopting and how are we mitigating these?

Have we evaluated our exposure to risks related to emerging technologies like AI-driven cyber attacks or the potential future impact of quantum computing on encryption?

How are we evaluating the risks associated with new technology investments such as AI, IoT, and cloud migration?

How are we managing and assessing the cyber risks associated with third parties, vendors, and our supply chain?

Do our partners and suppliers have risk-management processes that align with our cybersecurity standards?

How does our continuous risk profiling feed into our incident-response planning?

Do we have the agility to respond to new, unforeseen cyber threats in real time?

Are we staying ahead of regulatory changes and are these being incorporated into our risk-profiling process?

Do we have a plan to remain compliant with evolving data privacy regulations, especially considering new technologies we are adopting?

Are we investing in the right tools to ensure continuous risk monitoring, especially as we adopt new technologies?

How are we leveraging AI and/or automation to enhance our cybersecurity capabilities?

Is our cyber risk profile aligned with our broader business objectives, particularly in relation to digital transformation?

How are we leveraging AI, automation, and machine learning to enhance our cyber threat detection capabilities?

Are our cybersecurity tools adaptable enough to address future threats such as quantum computing and AI-driven attacks?

What percentage of our cybersecurity budget is allocated towards emerging technologies and is this sufficient to stay ahead of evolving threats?

How are we continuously training our employees to recognise and respond to the latest cybersecurity threats?

Do we have a plan to address the risks associated with insider threats and how is this integrated into our long-term cybersecurity strategy?

Are we investing in training programs that evolve with emerging threats and technologies?

What are the most significant emerging threats we expect to face in the next 5–10 years and how are we preparing for them now?

Is our cyber risk profile aligned with our broader business objectives and long-term growth plans?

How are we ensuring that our cybersecurity investment remains relevant and effective as our business grows and adopts new technologies?

Are we prepared to respond to new, unforeseen threats in real time and how do our current systems support this?

How are we integrating real-time threat intelligence into our incident-response planning to anticipate and mitigate future attacks?

APPENDIX D:
CYBER RISK APPETITE AND TOLERANCE STATEMENTS

Risk appetite statements serve as guiding principles that help organisations articulate their willingness to accept, manage, or avoid various types of risks in pursuit of their strategic objectives.

These statements establish clear boundaries for acceptable risk levels and inform decision-making processes across the enterprise. Any risks falling outside the defined appetite should prompt management action and, where appropriate, formal risk acceptance procedures such as documented exemptions.

Risk category	Cyber risk appetite statement
People	There is **no appetite** for a cyber event that may or will cause harm to an individual (as defined by the *Australian Privacy Act 1988*).
	There is **low risk appetite** for actions leading to significant stress, harm, or dissatisfaction among employees or the public due to poor implementation or inadequate support of ICT services.
	There is **no appetite** for insider threat leading to wilful and deliberate destruction of information assets. The information privacy principles aim to protect systems, data, and reputation from intentional harm and to uphold the highest standards of security and trust.
	There is a **low to moderate risk appetite** for insider threat as it is acknowledged that human error and minor unintentional errors can occur. It is therefore moderately accepted provided that such events are promptly detected and remediated without causing any harm.
	Repeated or severe accidental incidents that result in substantial data breaches, operational disruptions, compliance violations etc., that indicate systemic issues or inadequate training are unacceptable.
Reputational	The [company] is a trusted partner in the delivery and support of xxx. There is **low risk appetite** for major data breaches or failures in cybersecurity that lead to significant backlash and loss of trust.
	There is a **low risk** appetite for cyber events that impact multiple offices.
Operational	The [company] can tolerate moderate operational risks that may lead to temporary disruptions or delays. However, mechanisms must be in place to quickly restore operations and minimise the impacts.
	There is a **very low risk appetite** for unplanned downtime affecting xxx.
Financial	Some investment risk is acceptable to achieve long-term strategic goals like enhancing cybersecurity infrastructure if they deliver significant benefits and savings in the long run.
	There is **no appetite** for internal fraud events perpetrated by employees and contractors resulting from intent to defraud, misappropriation of assets, forgery, bribes, deliberate mismarking of positions, or theft.
	There is **no appetite** for external fraud events perpetrated by external threat actors resulting from intent to defraud the company, forgery, or theft.
	There is a **very low risk appetite** for incidents of fraud, collusion, or corruption resulting from inadequate vendor screening, lack of oversight, or insufficient internal controls which could lead to significant financial loss, reputational damage, or legal consequences.
Technology	There is high tolerance for the company to adopt technologies that enhance services, stakeholder engagement, and protection of data.
	There is **low risk appetite** for deploying untested technologies without adequate risk assessment and mitigation plans.
	There is **low risk appetite** for changes or failure in cybersecurity controls that negatively impact services resulting in loss of confidentiality, integrity, and availability of networked services and data.
Assets	There is **low risk appetite** for any incidents leading to unauthorised access, disclosure, alteration, or destruction of information assets which could lead to operational disruption, substantial harm to individuals, or reputational damage.

Risk category	Cyber risk appetite statement
Cloud services	There are strategic benefits of leveraging cloud services to enhance efficiency, scalability, and collaboration. The tolerance for cloud services is high, recognising that while cloud adoption can introduce potential vulnerabilities and compliance strategies, the benefits significantly contribute to operational goals.
	We mitigate these vulnerabilities through vendor risk assessments, contractual agreements, monitoring, and adherence to best practices in cloud security.
	There is **low risk appetite** for storing sensitive data on cloud platforms without proper encryption, access controls or failure to verify compliance with relevant regulations.
Compliance	The [company] has a **low risk appetite** for non-conformance to the xxx.
	There is low tolerance for ongoing/prolonged non-compliance to the requirements of the Privacy Act or an immature approach to cybersecurity risk management.
Third party	The [company] has high reliance on third-party products and services needed to enhance capability and capacity of services provided. The tolerance for supply chain risks is **low to moderate**, acknowledging that while reliance on external parties is necessary, it also introduces vulnerabilities. The [company] will partner with suppliers and vendors that are vetted and that demonstrate strong security practices and reliability.
	There is a **low risk appetite** for third-party non-compliance that has led or may lead to a cyber breach.

APPENDIX E:
RISK FRAMEWORK REVIEW

A robust risk framework is essential to safeguard an organisation against evolving cybersecurity threats, maintain regulatory compliance, and maintain operational resilience. The following checklist provides a structured approach for reviewing the effectiveness of your risk-management framework. Use this checklist to systematically assess each aspect, documenting your answers or observations in the space provided.

Now test the process: using a real-world event (ideally one that provides details on cause and impact), evaluate the risk event against your current risk matrix. Do the results indicate that it would be handled adequately? Are there any inconsistencies in likelihood and impact ratings? How easy was it to evaluate the hypothetical risk event using your matrix?

Checklist question	Your answer/observation
Are roles and responsibilities clearly defined in the risk framework?	
Do the requirements for risk evaluation cover all your obligations?	
Does the process cater for evaluating cybersecurity risks?	
Is cybersecurity risk appetite (what is / is not acceptable) clearly defined?	
Are factors that may influence risk acceptance and mitigation identified (e.g., is the reward greater than the potential risk)?	
Does the risk matrix align with your organisation's risk appetite?	
Are triggers for risk evaluation specified (e.g., projects, change control, organisational change, issues, near-miss events, incidents)?	
Does the risk framework specify how controls are selected?	
Does it specify how and when risks are updated?	

Checklist question	Your answer/observation
Is your risk matrix fit for purpose?	
Do the risk ratings and management decisions reflect the organisation's risk appetite?	
What does the matrix tell you about the organisation—is there a high tolerance for risk?	
Do existing information-security risks, based on impact and likelihood, accurately reflect reality?	
Have you evaluated the organisation's risk profile?	
Do the risks identified in the risk framework match the organisation's risk profile?	
Have you conducted any CTI activities and do the findings impact your risk assessments?	
Using a real-world event, does evaluation against your current risk matrix indicate it would be handled adequately?	

Checklist question	Your answer/observation
Are there inconsistencies in likelihood and impact ratings revealed through scenario testing?	
How easy was it to evaluate the scenario using your matrix?	

Now test the process: using a real-world event (ideally one that provides details on cause and impact), evaluate the risk event against your current risk matrix. Do the results indicate that it would be handled adequately? Are there any inconsistencies in likelihood and impact ratings? How easy was it to evaluate the hypothetical risk event using your matrix?

APPENDIX F:

OKRS

The following table provides some sample OKRs that should be adapted for your needs and capabilities—remember that objectives also need to be *achievable* and *realistic (SMART)*.

Objective	Key Result
Strengthen collaboration with key stakeholders	Develop and share 10 tailored threat reports with cross-functional teams (e.g., incident response, legal, PR) within the next quarter.
	Host quarterly CTI briefings for senior leadership and operational teams, achieving 90% attendance.
	Establish partnerships with 3 external intelligence-sharing platforms or industry ISACs (information sharing and analysis centres) within 6 months.
Enhance threat visibility across the organisation	Increase the number of threat feeds and intelligence by 30% in the next quarter.
	Achieve a 95% correlation rate between threat intelligence data and internal threat detection alerts.
	Expand coverage to identify at least 90% of targeted threat vectors relevant to the organisation's sector.
Improve the quality and actionability of threat intelligence	Reduce the average time to validate and prioritise threat intelligence reports by 20% within 6 months.
	Achieve a 90% satisfaction rate from security analysts on the relevance of CTI reports provided.
	Decrease the percentage of false positives in threat alerts to less than 10%.
Optimise threat hunting and analysis capabilities	Conduct 12 targeted threat-hunting exercises in high-risk areas over the next year.
	Identify and neutralise at least 10 previously unknown vulnerabilities or attack patterns during these exercises.
	Implement an automated CTI analysis tool that reduces manual analysis efforts by 40%.
Measure and demonstrate the value of CTI programs	Track and report the mitigation of at least 20 high-risk threats using CTI insights over 6 months.
	Quantify and present an estimated ROI for CTI activities, demonstrating at least 3x cost savings from prevented incidents.
	Conduct biannual stakeholder surveys, achieving at least an 85% positive perception of CTI's effectiveness.
Enhance proactive threat defence	Reduce the average detection-to-mitigation time for priority threats by 25% within 6 months.
	Implement a proactive threat intelligence campaign addressing 5 emerging threats over the next quarter.
	Train 100% of the incident-response team in interpreting and applying CTI insights effectively.
Improve threat intelligence automation	Deploy automation tools to process 80% of incoming threat data within 90 days.
	Integrate CTI automation with existing SIEM and SOAR tools to automate at least 50% of threat triage processes.
	Conduct a post-implementation review showing a 30% improvement in efficiency.
Enhance strategic intelligence for long-term risk management	Publish 3 strategic CTI reports on long-term threat trends and geopolitical risks within the next year.
	Present intelligence on evolving risks to the board biannually, with actionable recommendations in each session.
	Develop risk scenarios for 5 major adversary groups to inform strategic defences.

Select the most relevant OKRs for your organisation, considering its risk appetite and risk profile. What do you want to manage well to give the board and other key stakeholders assurance that the organisation is secure?

Once you've established the right OKRs for your organisation, the next step is to work out how you will gain insight into the KRs. What do you need to do and what do you need to monitor to obtain the required results?

APPENDIX G:

GLOBAL REGULATORY MAPPING FRAMEWORK

Privacy and data-protection regulations

Legislation/standard	Jurisdiction	Key requirements	How requirements are met by ISMS/AIMS
GDPR (General Data Protection Regulation)	European Union	• Right to be forgotten • Data portability • Privacy by design • DPIA requirements • 72-hour breach notification	• Privacy management plan with EU-specific procedures • Data subject rights management system • Privacy impact assessment framework • Automated breach detection and notification
Australian Privacy Act 1988 (amended 2024)	Australia	• Secure handling of personal data • Enhanced data breach notifications • Stricter consent requirements	• Privacy management plan with Australian-specific procedures • Data breach response plan incorporates the Office of the Australian Information Commissioners (OAIC) reporting requirements
CCPA/CPRA (California Consumer Privacy Act)	California, USA	• Consumer rights to know, delete, opt out • Data minimisation requirements • Third-party data-sharing disclosure	• Consumer rights management system • Data inventory and mapping procedures • Vendor data-sharing agreements

PDPL (Personal Data Protection Law)	Saudi Arabia	• Data localisation requirements • Cross-border transfer restrictions • Mandatory appointment of a data protection officer	• Data residency compliance framework • Cross-border transfer impact assessment • Local data protection officer designation
PDPA (Personal Data Protection Act)	Singapore	• Consent management • Data breach notification (3 days) • Data protection officer requirements	• Consent management platform • Automated breach notification to PDPC • DPO governance framework

Cybersecurity regulations

Legislation/standard	Jurisdiction	Key requirements	How requirements are met by ISMS
NIS2 Directive	EU	• Risk-management measures • Incident reporting (24 hours) • Supply chain security • Business continuity planning	• Comprehensive risk management framework • Automated incident detection and reporting • Third-party risk assessment procedures • Business continuity management system
Australian Cybersecurity Act 2024	Australia	• Mandatory ransomware reporting • Critical infrastructure protection • Government contractor requirements	• Ransomware response plan with government reporting • Critical infrastructure risk assessment • Enhanced security controls for government contracts
Cybersecurity Enhancement Act	USA	• Federal contractor cybersecurity • NIST framework implementation • Supply chain risk management	• NIST CSF alignment • Federal contractor compliance procedures • Supply chain security assessment
Network and Cybersecurity Act (NCA)	Saudi Arabia	• Cybersecurity controls implementation • Incident reporting to NCA • Regular security assessments	• NCA-compliant security control framework • Incident response with NCA notification • Periodic security assessment schedule
Cybersecurity Act 2018 (CSA)	Singapore	• Critical information infrastructure (CII) protection • Cybersecurity incident reporting • Regular penetration testing	• CII owner security measures • CSA incident reporting procedures • Penetration testing program

AI governance regulations

Legislation/standard	Jurisdiction	Key requirements	How requirements are met by AIMS
EU AI Act	EU	• Risk-based AI classification • High-risk AI system requirements • Prohibited AI practices • CE marking for AI systems • Fundamental rights impact assessment	• AI risk classification framework • High-risk AI governance procedures • Prohibited use case policies • AI system documentation and marking • Fundamental rights impact assessment
Australian AI Ethics Framework	Australia	• Human-centred AI design • Fairness and non-discrimination • Privacy protection and security • Reliability and safety • Transparency and explainability • Contestability and accountability	• AI ethics committee establishment • Bias detection and mitigation procedures • AI system transparency requirements • Human oversight and intervention protocols
NIST AI Risk Management Framework	USA	• AI risk governance • AI system lifecycle management • Trustworthy AI characteristics • Continuous monitoring and improvement	• AI risk-management system • AI lifecycle governance procedures • Trustworthiness assessment framework • AI performance monitoring system
Model AI Governance Framework	Singapore	• AI governance structure • Risk management and internal controls • Human involvement in AI decisions • Operations management • Stakeholder interaction and communication	• AI governance committee • AI risk-assessment and control framework • Human-in-the-loop requirements • AI operations management procedures • Stakeholder communication protocols

Industry-specific regulations

Sector	Key regulations	Jurisdiction	Requirements	ISMS/AIMS integration
Financial services	• APRA CPS 234 • PCI DSS • SOX • MiFID II • Basel III	Australia, global, USA, EU	• Information security capability • Payment card data protection • Financial reporting controls • Algorithmic trading oversight • Operational risk management	• Financial services risk framework • Payment security controls • Financial reporting ISMS • AI trading system governance • Operational resilience planning
Healthcare	• HIPAA • TGA • MDR • FDA AI/ML guidance	USA, Australia, EU	• Protected health information security • Medical device cybersecurity • AI/ML medical device regulation • Clinical data protection	• Healthcare data protection framework • Medical device security controls • AI/ML clinical validation procedures • Patient data governance
Critical infrastructure	• Critical Infrastructure Protection Act • NIS2 • NERC CIP • Cybersecurity Framework	Australia, EU, USA	• Asset identification and protection • Incident response and recovery • Supply chain security • Continuous monitoring	• Critical-asset protection framework • Infrastructure incident response • Supply chain risk management • Continuous security monitoring

APPENDIX H:

EXAMPLE PESTEL ANALYSIS

The following table demonstrates the use case for a cyber-centric PESTEL analysis.

	Influence/impact on cybersecurity	Strategy
P	Government procurement contracts now specify ISO 27001 requirements for organisations providing managed services. New CEO requires regular reports on cyber risks.	Implement ISO 27001. Design and implement monthly cybersecurity metrics and reporting.
E	Budget restrictions indicate we need to refine and restrict the adoption of new IT systems.	Evaluate current systems and technical controls, and identify ROI.
S	Clients are increasingly aware of cybersecurity events and will not tolerate repeated or similar data breach events. Increased likelihood of ransomware attacks against similar organisations.	Identify historical/industry data-breach events and use to evaluate the effectiveness of the cybersecurity strategy and incident-response processes. Evaluate the cyber resiliency plan (ability to detect, respond, and recover from a ransomware attack).
T	CTO is encouraging the adoption of AI across the organisation. We need to watch quantum computing advancements.	Define the AI strategy, and supporting policies and procedures. Evaluate how quantum computing will impact the effectiveness of current encryption technologies used by the organisation.
E	Increased numbers of staff and contractors are working from home.	Implement DLP tools and MFA.
L	New cybersecurity regulations are in place requiring the organisation to report ransomware payments.	Develop a ransomware plan.

APPENDIX I:

INDUSTRY PROFILING

Establishing context for any cybersecurity program should be grounded in the specific industry sector. By covering the 5 key sectors below as examples—finance, healthcare, the autonomous and electric vehicle industry, mining & natural resources, and education & research institutions—we can demonstrate how motives and threats are shaped by the products and services each provides.

FINANCE

Financial institutions face heightened risk compared to other sectors because the nature of their business, managing large volumes of monetary transactions and sensitive client data, makes them especially attractive targets. The direct access to funds, personal information, and confidential account details creates numerous opportunities for cybercriminals to perpetrate financial fraud, facilitate money laundering, and exploit vulnerabilities for illegal gain. As a result, the financial services industry is continuously exposed to sophisticated schemes and persistent threats seeking to capitalise on these unique risks.

Motive	Unauthorised access to accounts for financial gain
	Copying or extracting personal data for the purpose of identity theft (financial gain)
	Access to confidential business information for personal gain (insider trading)
Threat actor	Malicious insider, hacker, cybercriminal, advanced persistent threat (APT)
Attack vectors (method or pathway used to exploit vulnerabilities)	Business email compromise (known as BEC): phishing and impersonation scams targeting financial transactions
	Banking trojans and ransomware: malware designed to steal banking credentials or lock financial systems
	Supply chain attacks: targeting third-party payment processors or fintech apps to gain access to customer data
	Insider threat: employees exploiting access to sensitive systems

HEALTHCARE (HOSPITALS, PHARMACEUTICAL, TELEMEDICINE)

Healthcare organisations manage critical and sensitive patient data, making them attractive targets for ransomware attacks and data breaches. The urgent nature and dependency on technology uptime mean ransomware attacks are particularly successful in this sector and an attractive disruptor for cybercriminals.

Motive	Copying or extracting personal data for the purpose of identity theft (financial gain) Dark web data sales: stolen patient records are more valuable than credit card data on illicit markets Causing a disruptive event in protest or for financial gain
Threat actor	Malicious insider, hacker, cybercriminal, APT
Attack vectors	Ransomware attacks: disrupting hospital operations, forcing institutions to pay to restore access Medical device exploitation: IoT-connected devices (e.g., pacemakers, insulin pumps) as potential attack vectors

AUTONOMOUS AND ELECTRIC VEHICLE INDUSTRY

Innovative organisations such as the EV industry are highly dependent upon software, AI, and integrated infrastructure, making them vulnerable to attacks that target IP related to self-driving algorithms and EV battery technology.

Motive	To sabotage or disrupt projects, products or services Copying or extracting IP for personal gain
Threat actor	Malicious insider, hacker, cybercriminal, APT
Attack vectors	IP theft and espionage: nation-state actors targeting AI-driven vehicle systems and proprietary software Remote hijacking: exploiting vulnerabilities in connected car APIs or charging-station networks Supply chain attacks: targeting chip manufacturers or EV software suppliers to implant back doors

MINING & NATURAL RESOURCES

This sector has heavy reliance on industrial control systems (ICS) and OT, which are often outdated and vulnerable. Valuable geological survey data and IP on resource deposits attract espionage and cyber theft. Mining operations are critical infrastructure, making them targets for ransomware and geopolitical cyber threats.

Motive	Disruption of critical infrastructure for political or financial gain
	Theft of proprietary data related to resource extraction
Threat actor	Malicious insider, hacker, cybercriminal, APT
Attack vectors	Sabotage: shutting down mining operations by damaging physical equipment and IoT devices
	Ransomware attacks on ICS/OT: shutting down mining operations by locking out control systems
	Nation-state espionage: stealing data on new resource discoveries and extraction technologies
	IoT device exploitation: attacks on smart sensors, autonomous haul trucks, and automated drilling equipment
	Supply chain vulnerabilities: cybercriminals infiltrating third-party contractors (equipment manufacturers, logistics)
	Insider threat: employees exploiting access to sensitive systems to obtain proprietary data to benefit themselves or a competitor (espionage)

EDUCATION & RESEARCH INSTITUTIONS

Education and research institution store vast amounts of student, staff, and research data (including government-funded projects). They often have weak security policies and large, decentralised networks, making them easy targets. High-value research (e.g., medical advancements, technology patents) is attractive for nation-state cyber espionage.

Motive	Stealing PII of students and staff, as well as valuable research data
	Financial gain: extorting institutions through ransomware attacks or selling stolen data on the dark web
	Espionage: accessing cutting-edge research for competitive or national advantage
Threat actor	Malicious insider, hacker, cybercriminal, APT
Attack vectors	Phishing and credential theft: students and staff targeted to access academic networks
	Ransomware attacks: schools and universities often lack budget for cybersecurity, making them vulnerable
	Data breaches: theft of PII, financial aid records, and IP from research institutions
	DDoS attacks: disrupting virtual classrooms, learning platforms, and university operations

APPENDIX J:
TOOLS FOR NETWORK AND INFORMATION FLOW MAPPING

Depending on the organisation's needs, different tools can be used to map and analyse networks and information flows as follows:

1. Network discovery & visualisation tools

- SolarWinds Network Topology Mapper—automatically discovers and visualises network topology
- Microsoft Visio—manual network diagramming tool for structured mapping
- Lucidchart—cloud-based diagramming tool for mapping infrastructure

2. Security & attack-surface mapping tools

- Tenable.sc/Nessus—conducts vulnerability scanning and maps exposures in the network
- Censys/Shodan—identifies externally facing assets and possible misconfigurations
- AttackIQ/SafeBreach—simulates attacker behaviour and maps security gaps

3. Data flow & dependency mapping tools

- AWS VPC Flow Logs / Azure Network Watcher—provides network traffic flow insights in cloud environments
- MISP (Malware Information Sharing Platform)—helps track data movement related to cyber threats

- Splunk / Elastic Security—correlates logs and tracks how data moves across systems

4. Cloud & hybrid infrastructure mapping tools

- Cisco Meraki Dashboard—monitors network devices and cloud-managed security
- CloudMapper (for AWS)—visualises AWS environments and identifies misconfigurations
- Google Cloud Network Intelligence Center—provides network visibility and diagnostics

5. Asset & third-party risk-mapping tools

- ServiceNow CMDB—tracks IT assets, their configurations, and interdependencies
- Armis—maps IoT and unmanaged devices for visibility in OT/ICS environments
- RiskRecon—assesses third-party cybersecurity risks by mapping vendor networks

APPENDIX K:

THREAT AND CAUSE EVENTS

Threat name	Cause event	Associated event
Unplanned downtime / loss of availability	Third party / supply chain suffers an event	• Non-compliant/immature controls • Poor procurement processes
	Third party / supply chain causes an event	
	Unauthorised/unplanned changes	• Internal/external non-compliance (deliberate or accidental) • Inadequate system segregation
	Infrastructure failure	• Software/hardware failure
	Technical failure	• Software/hardware failure • Legacy system
	Loss/damage to physical assets	• Theft • Accidental/deliberate damage • Environmental damage
	Poor vulnerability management	• Inadequate monitoring activities • Inadequate CTI activities • Inadequate patch management
Organisational threat	Third party non-compliance / failure	• Poor procurement / validation of capabilities • Poor ongoing management / oversight of third party
	Insider threat—human action/inaction	• Inadequate training/awareness • Disgruntled employee • Loss or damage to an asset
	Fraud/theft	• Lack of segregation • Inadequate onboarding/offboarding processes • Inadequate DLP capabilities • Poor procurement practices
	Model stealing / loss of IP	• Model inversion (AI) • Theft of trained AI models • Inadequate DLP capabilities

Threat name	Cause event	Associated event
	Unauthorised access	• Poor access and identity management • Inadequate system monitoring capabilities • Poor vulnerability management
	Poor governance/oversight	• Inadequate training/awareness • Lack of accountability • Lack of staff/resources • No/immature cybersecurity plan • Inadequate asset management • Inadequate risk management • Poor change control
Hacker / state sponsored attacker	Phishing / spear phishing	• Inadequate training/awareness • Inadequate prevention controls
	Social engineering	• Inadequate training/awareness • Poor procurement practices
	Denial of service / ransomware	• Poor vulnerability management
	Unauthorised access to AI system	• Data and model poisoning (injecting false information to mislead AI systems) • Evasion attacks (crafting inputs to deceive AI algorithms). • Data manipulation (techniques to manipulate models)
	Immature CTI activities across agencies	• Inadequate training/awareness • Poor governance/oversight of cybersecurity program (per agency)
	Inadequate detection mechanisms	• Inadequate system logs / monitoring capabilities

APPENDIX L:

LIKELIHOOD—BASED ON CONFIDENCE IN CONTROLS

Likelihood	Confidence in controls	CTI
Rare	All baseline and key controls are in place and operational at level 4 maturity or above. There is a high level of control automation. KRIs are tracked and reported regularly.	Requires highly skilled and funded professional threat actor. Insider threat is deliberate.
Unlikely	Appropriate controls are in place and audited. Controls enable immediate identification and containment of occurrence.	Requires highly skilled and motivated threat actor. Insider threat is deliberate.
Possible	Controls are mostly effective. All baseline controls are in place and operational at level 3 or above. Some key controls have been implemented.	Threat actor is highly motivated. Insider threat is deliberate/accidental. Targeted attack is for personal gain, ideology, or revenge.
Likely	Controls are partially effective. No key controls have been defined. Majority of baseline controls have been implemented and are operational at level 2 or above.	Minimal skill requirements are needed to exploit the vulnerabilities. Insider threat is accidental. Is motivated by social recognition, fun, challenge.
Almost certain	Controls are ineffective. Reporting mechanisms are ad hoc, reactive. Control maturity is less than level 3.	There are active CTI reports and activity related to the assets used by government.

APPENDIX M:

LIKELIHOOD—BASED ON FREQUENCY

Likelihood	Frequency of occurrence (known or expected)	Incidents
Rare	Is a plausible event maybe 1/50 years.	No prior events have occurred.
Unlikely	May occur every 5–10 years.	Events have occurred within the past 10 years.
Possible	Is possible that event will occur with notable increase in events occurring elsewhere.	Several events have occurred in the past 2–5 years.
Likely	Is expected to occur at least once a year. There are active threat actors known to be exploiting vulnerabilities.	Common occurrence and several events have occurred in the past 6–12 months.
Almost certain	Is a regular occurrence, daily or weekly. There are active threat actors known to be exploiting vulnerabilities.	Common occurrence and several events have occurred in the past 6 months.

APPENDIX N:

CONSEQUENCE RATINGS

Consequence	Scope	IT services	Data compromise
Insignificant	Event impacts a department or team.	No direct impact to IT services	Unclassified data is compromised.
Minor	Event impacts one department but can be easily overcome.	Limited disruption to operations; minor delays to SLA that are manageable	Small group of impacted individuals' data is compromised.
Moderate	Event impacts one department and may result in degraded services. SLA cannot be met.	Possible unplanned downtime or degradation of IT services SLA cannot be met.	May lead to harm to an individual (privacy).
Major	Event impacts multiple departments.	Prolonged system downtime or degradation of IT services but within recovery time and recovery point objectives (RTO/RPO)	Vast amount of personal and in-confidence data is compromised.
Critical	Event impacts all departments. Impact can be long-lasting.	Significant system downtime or degradation of IT services exceeding RPO/RTO	Full system is compromised across multiple departments.

APPENDIX O:
VELOCITY OF A CYBER RISK EVENT

In evaluating and responding to cyber risks, understanding the velocity (or the speed at which a cyber security event unfolds) is crucial. The rate at which a threat develops and propagates can dramatically influence the effectiveness of mitigation strategies, the allocation of resources, and the prioritisation of incident response.

Some incidents emerge slowly, providing time to react, while others escalate rapidly, leaving little opportunity for containment or recovery. Recognising the velocity of a cyber security event enables organisations to tailor their preparedness and response measures, minimising potential damage and ensuring a more resilient posture.

The table below offers examples to help refine impact statements and highlight that even slow-moving events can have severe consequences once they gain momentum.

It categorises cyber risk events by velocity, with descriptions and examples for each level. This helps organisations better anticipate how quickly threats may arise and plan effective responses.

Velocity	Description	Examples
1 Extremely slow	Threat takes a long time to develop and spread.	Long-term vulnerabilities that require significant time and effort to exploit. Rarely targeted.
		Legacy systems. Obsolete protocol attacks.
2 Very slow	Threat needs time to propagate or requires physical access.	Rare vulnerability that requires specific conditions to exploit. Attacker aims for long term.
		State-sponsored attacks. APTs.
3 Slow	Threats emerge and spread at a gradual rate.	Low-priority vulnerabilities that are not widely targeted or exploited.
		Firmware exploits.
4 Moderately slow	Threats emerge and spread at a gradual rate.	Exploits that require detailed knowledge and extensive setup to execute.
		Logic bombs / malware that activates under specific conditions. May lie dormant.
5 Moderate	Threats have a balanced rate of emergence and propagation.	Standard threats that are known but require moderate effort to exploit.
		Insider threats. Social engineering.
6 Moderately fast	Threat can spread at a faster than average rate.	Common exploits that can be automated and executed with moderate ease.
		Brute force attacks. Stolen credentials.
7 Fast	Threat can spread quickly and can cause harm in a short period.	Common vulnerabilities that are easy to exploit and can propagate rapidly.
		SQL injection attacks. Phishing campaigns. Supply chain attacks.
8 Very fast	Threats propagate and cause harm very quickly.	Attacks that can disrupt services almost immediately.
		Ransomware and DDoS attacks.
9 Extremely fast	Threats spread and cause significant harm almost instantaneously.	Attacks that can infect thousands of systems fast.
		Worms like WannaCry.
10 Instantaneous	Threats cause immediate and widespread harm upon their emergence.	Zero-day exploits that are used immediately before any patches or defences can be applied.
		Supply chain attacks (e.g. SolarWinds) – compromising trusted software updates.

APPENDIX P:

THREAT HUNTING—KEY STEPS

The steps outlined below present a streamlined and high-level approach, and serve to illustrate the essential connection between tactical CTI and effective risk profiling. By framing tactical CTI as an actionable input into the risk-management process, this appendix demonstrates how immediate threat insights can directly inform the identification, assessment, and prioritisation of organisational risks. Although real-world scenarios often require greater nuance and complexity, this simplified view offers a practical foundation for understanding how intelligence-driven action can enhance your overall risk posture.

Define the hunt hypothesis

- Formulate a hypothesis based on the latest CTI and recent attack trends.

Example: "Given increased phishing campaigns targeting our sector, attackers may attempt to compromise HR systems via spoofed emails."

Scope and prioritise assets

- Identify critical systems, networks, and data relevant to your organisation's business operations.
- Use CTI insights to prioritise the assets most likely to be targeted.

Example: "Prioritise healthcare databases if CTI highlights ransomware activity in the healthcare sector."

Data collection and analysis

- Gather information from logs, network traffic, endpoint monitoring, and other detection systems.
- Leverage IOCs and suspicious patterns identified in CTI reports.

Hunt execution

- Search for malicious activity using behavioural analysis and cross-reference findings with known attacker TTPs and IOCs.

Example: "Detect lateral movement in the network using patterns aligned with attacker behaviour."

Detection and investigation

- Investigate anomalies to determine whether they indicate genuine threats.
- Use CTI to distinguish between benign incidents and active compromises.

Example: "Validate unexpected login attempts with global brute-force trends seen in CTI."

Mitigation and remediation

- If threats are confirmed, enact incident-response processes like isolating impacted systems or deploying patches.
- Enhance security controls (e.g., firewalls, email filters) based on lessons learned.

Documentation and reporting

- Record findings, steps taken, and outcomes.
- Share insights with relevant teams to update threat profiles and defensive strategies.

Continuous improvement

- Use threat-hunting results to refine your risk profile and CTI program.
- Adapt organisational strategies to address newly observed vulnerabilities and attacker trends.

APPENDIX Q:
TACTICAL CT: PRACTICAL ACTIVITIES YOU CAN DO NOW

Outline your current tactical CTI capabilities and identify areas for improvement:

What threat feeds are you subscribed to?

There are various resources for organisations to gather intelligence on known threats, such as MITRE ATT&CK Updates, which "is a globally accessible knowledge base of adversary tactics and techniques based on real-world observations. The ATT&CK knowledge base is used as a foundation for the development of specific threat models and methodologies in the private sector, in government, and in the cybersecurity product and service community."

https://attack.mitre.org

What additional intelligence do you have access to?

For example, are you signed up to industry forums and special interest groups on Slack/Discord and LinkedIn?

What threat monitoring tools do you already have access to?

If you're on an M365 licence, review your Microsoft security portal. Check for threat intelligence reports under Defender for Endpoint (free with some M365 plans). It provides a curated threat feed, indicators of compromise (IOCs), and attack trends.

Microsoft Sentinel (free tier) is a cloud-based security information and event management (SIEM) with free connectors for threat intelligence.

Do you understand your key threat actors?

It's necessary to understand which adversaries are most likely to target your organisation and why, so you can learn their behaviours and tactics. "Know thine enemy!"

APPENDIX R:
AI RISK-PROFILING TOOL

As AI becomes increasingly embedded within organisational processes, it is essential to adopt a systematic approach to identifying, assessing, and managing the unique risks it introduces. This AI risk-profiling tool has been developed to provide a structured framework for evaluating potential exposures associated with AI adoption and operations. By mapping AI usage across the business, assessing oversight and ownership, and scrutinising data and model integrity, the tool equips stakeholders with the insights necessary to make informed decisions, strengthen controls, and uphold compliance obligations.

Category	Key questions & considerations	Indicators of elevated risk	Rating (R/A/G)
AI usage mapping	Have you mapped where AI is used across the business?	Shadow AI in business units	
	Is there visibility into who owns or manages each system?	No central oversight or ownership of AI initiatives	
Data & model integrity	What data is being used to train or operate AI systems?	Poor data quality Unknown data sources	
	Is the data accurate, representative, and governed?	No data validation before model training	
Risk & ethics governance	Are there policies in place for AI ethics, transparency, and decision accountability?	No documented policy on AI use	
	Is there a review board or sign-off process?	AI used in decision-making without human oversight	
Security & adversarial risk	Are models protected against manipulation (e.g., prompt injection, poisoning)?	No model monitoring No drift detection	
	Are outputs monitored for drift or misuse?	Lack of adversarial threat awareness or testing	
Business impact & resilience	Has the business assessed the impacts of AI failure or incorrect decisions?	No contingency planning for AI errors	
	Are there incident-response plans for AI-driven services?	High dependency on opaque AI systems	

Guide:

- For each category, answer the questions honestly and assign red/amber/green status.
- Use the results to highlight priority focus areas in your AI governance and cyber risk management processes.
- Where multiple red or amber ratings exist, you likely need:
- Stronger risk-profiling and risk-ownership structures
- Updates to governance frameworks
- Additional monitoring, review, or ethical controls.

APPENDIX S:
AI RISK AND PERFORMANCE METRICS

The following table provides some example metrics that could be considered within an AIIA.

Category	Metric	Purpose	Who should monitor
Model performance	Model accuracy / error rate	Measures how well the AI is performing its intended task	AI owner / data science lead
	Model drift detection frequency	Detects frequency of changes in model behaviour due to changing data inputs	Data science lead
	False positive/negative rate	Indicates potential harm from incorrect decisions	Risk officer / audit team
Operational risk	System downtime/availability	Tracks AI system availability and continuity	IT / ops
	Override rate (human intervention)	Tracks how often humans override AI outputs	Product owner / compliance
	Escalated AI failures (per quarter)	Tracks how many times AI failure required cross-functional escalation	CIO / risk committee
Ethics & compliance	Bias indicators (e.g., disparate outcomes by demographic group)	Detects unintentional bias that could breach laws or trust	Ethics board / legal
	Transparency score (based on explainability audit)	Measures how well decisions can be explained or understood	AI governance committee
	Personal data processed (% classified data used)	Tracks PII or sensitive data usage and related obligations	DPO / privacy team
Governance	AI risk-register updates (Y/N each quarter)	Ensures AI-specific risks are documented and managed	Risk owner / GRC lead
	AIIAs completed	Tracks the number of AI projects reviewed through an AIIA process	Project management office
	Number of AI-related incidents or breaches	Keeps visibility of near misses, misuses, and breaches	Security/risk officer
Culture & readiness	Staff trained on AI acceptable use policy (% coverage)	Gauges internal readiness and guardrail awareness	HR / risk
	Staff confidence score (via survey)	Monitors culture and workforce comfort in using or challenging AI outputs	People & culture / strategy
	AI tools in use (number across organisation or per department)	Helps track AI tool sprawl and shadow AI risks	IT / architecture

APPENDIX T:
AI RISK OVERSIGHT FOR EXECUTIVES

The following table demonstrates how metrics and associated AI risks could be reported to executives each quarter.

AI governance dashboard—exec summary report for Q2				
Metric	**Target**	**Current**	**Status**	**Trend**
Model accuracy (top 5 AI systems)	≥ 95%	91%	◆ Warning	🔽 Down
AIIAs completed	100% of new use	85%	◆ Warning	→ Stable
AI-related incidents	0	2	● Alert	🔼 Up
Staff trained on AI acceptable use	100%	64%	◆ Warning	🔼 Improving
Bias detection score (loan model)	No disparities	Moderate	◆ Warning	→ Stable
Model drift events detected	≤ 1/quarter	0	● OK	→ Stable
Transparency audit score	≥ 4 / 5	3	◆ Warning	→ Stable
Key risks				
Inadequate explainability in decision-support AI				
Rising incident rate in customer-facing automation				
Mitigation				
Initiate post-incident reviews.				
Prioritise staff training.				
Refresh bias testing on financial models.				

APPENDIX U:
INCIDENT-RESPONSE INTEGRATION TEMPLATE

This template provides a structured approach to documenting and managing cybersecurity incidents, risk assessments, and ongoing response activities. It is designed to help organisations track incidents, analyse root causes, assess changes to their risk profiles, and validate that mitigation actions are implemented effectively.

In addition to incident recording, it guides regular review of response plans and preparedness activities, supporting continuous improvement of security posture. Users should complete each section as incidents occur and update the status of mitigation and review actions according to the prescribed frequency, ensuring accountability and alignment with evolving risks.

Incident-response plan alignment with risk profile

Risk type	Corresponding incident-response playbook	Last reviewed	Test frequency	Responsible team
Ransomware	Containment & comms	Feb 2025	Twice yearly	Security ops / comms
Insider threat	Insider threat playbook	Mar 2025	Annual	HR / security lead

Only include playbooks tied to highest risk areas and those with high velocity/impact.

Human factor considerations

Factor	Incident-response mitigation strategy	Training frequency	Notes
Phishing-prone staff	Adaptive phishing simulations	Quarterly	Tailored based on click-rate risk
Insider risk roles	Behaviour monitoring Whistleblower hotline	Annually	Must align with DISP requirements (if applicable)

If humans are the biggest risk, they must also be your first line of detection.

Communication triggers based on risk profile

Event type	Internal notification time	External notification (regulators, public)	Comms owner
Personal data breach (PII)	30 minutes	Within 72 hrs (as per OAIC)	Privacy officer
Ransomware affecting operations	Immediate	Case by case	Crisis comms lead

Communication cadence and tone should reflect both legal and reputational risks.

Lessons learned / profile update

Incident	Date	Root cause	Risk profile update?	Mitigation action
Phishing + credential theft	Jan 2025	MFA gap	Yes—Likelihood ↑	Enforce MFA on all endpoints

Post-incident reviews must explicitly ask: "Does this change our risk profile?"

Review & audit schedule

Activity	Frequency	Responsible role	Notes
Incident response plan review	Every 6 months	CISO/ISM	Cross-check with updated risk register
Tabletop simulation	Annual (minimum)	Cyber risk committee	Prioritise high-velocity threats

Scenarios should mirror real threats, not generic ones, based on your top 5 risks.

APPENDIX V:
BUILDING SCENARIOS AND STRATEGIES

With a clear definition of the business processes, their dependencies and their risk profiles, we can start to build a picture of the most relevant threats to the organisation's strategic objectives.

To make your risk-assessment process practical and impactful, start by reflecting on your organisation's unique context and vulnerabilities using the results of your PESTEL analysis. Ask yourself: Based on what you know, which disruptive scenarios are most plausible for your business.

- **Identify relevant scenarios**: list possible disruptive events (e.g., cyber attacks, supplier failures, natural disasters) that are most likely to affect your organisation. Don't guess—use data and insights from your PESTEL analysis and any previous incidents.
- **Engage stakeholders**: circulate your preliminary list of risk scenarios among key internal stakeholders. Solicit feedback to verify your list is comprehensive and all crucial concerns are covered.
- **Leverage historical data:** gather internal metrics and external case studies to inform your scenarios. If your organisation lacks direct experience, research industry incidents and post-incident reports to enhance your planning.

Dependency	Event	Relevant scenario
People	Pandemic	COVID
	Loss of staff	Sickness/death of a key person
		High turnover of staff
		Culture / reputational damage (bad culture)
Facilities / geographical location	Loss of facilities	Location destroyed
		Partially damaged facilities
		Malfunctioning utilities
	Terrorism	Weapon on site
		Bomb threats
Technology/data	Loss of IT	Ransomware—full loss of systems
		Partial loss (single system unavailable)
		Unplanned IT downtime
	Loss of data/IP	Hacker (compromised IT systems)
		Unauthorised access / disclosure of information
		Insider threat (espionage)
Financial	Fraud	Insider threat (theft or alteration of records)
		Falsified accounts
Supply chain	Loss of supplier/provider (disruption to supply chain)	Loss / reduction of service from key supplier
		Poor quality = reputational damage
		Suppliers in financial difficulty
Reputational	Data breach / negative PR event	Negative social media / news

- **Document and prioritise**: maintain this as a high-level overview, prioritising events by likelihood and potential impact. Keep the focus on the business: its operations, dependencies, and clientele.

The objectives of this step is to clarify the most relevant strategies and plans needed to prevent or manage the relevant scenarios.

- **Develop mitigation strategies**: based on your top risks, outline or refresh core plans such as:
  - Pandemic response plan
  - Cybersecurity strategy
  - Supplier management plan
  - Emergency management plan
  - Business continuity strategy
- **Continuous improvement:** integrate new learnings and stakeholder input regularly, using updates to refine your mitigation strategies and scenario list.

By following these actions, you'll gain a clear, business-focused view of your risk landscape, ensuring your continuity and cybersecurity plans are both relevant and robust.

APPENDIX W:
BOARD AND EXECUTIVE CYBERSECURITY REPORT TEMPLATE

EXECUTIVE SUMMARY

Current cybersecurity maturity level	Basic/developing/advanced
Key findings	Summarise key insights from the control maturity assessment.
Top 3–5 critical risks	Briefly list major threats and vulnerabilities.
Business impact	Describe how risks affect operations, revenue, compliance, reputation.
Recommended actions	Summarise urgent steps and long-term initiatives.

Cybersecurity maturity & benchmarking

This section of the report provides a structured and comprehensive overview of your organisation's current cybersecurity maturity. It leverages industry-standard frameworks and benchmarks to offer actionable insights, highlight critical risks, and outline clear next steps to strengthen your security posture. Each topic covered below plays a crucial role in ensuring holistic understanding and effective mitigation of cyber threats.

Gap analysis summary

- This summary is a concise overview of the most critical gaps between your current controls and those recommended by industry standards. It identifies priority areas that require immediate or focused attention.

Current state versus industry standards

- This subsection compares your organisation's cybersecurity maturity level against recognised industry frameworks such as

NIST CSF, ISO 27001, or CIS Controls. It helps identify how your control environment stacks up against best practices and regulatory requirements.

Maturity score

- Here, a quantified score is presented based on the selected frameworks. This gives a clear, objective snapshot of overall maturity and is useful for tracking progress over time.

Comparison with industry peers

- This subsection benchmarks your organisation's security capability against similar organisations, using visual aids where possible. Such comparisons can help prioritise areas for investment and improvement.

Desired target state

- Based on the findings against industry standards, regulatory requirements and industry peers, here you clearly define the desired target state for your organisation. Link to organisational goals and objectives where possible, and link to the organisation's risk profile and stakeholder expectations to create the desired target state and timeframe for achieving it.

KEY RISKS & BUSINESS IMPACT

This section identifies and categorises the most significant threats and vulnerabilities facing the organisation, and articulates their potential consequences.

Risk category

- This provides a breakdown of the types of risks (such as data breaches, insider threats, system vulnerabilities) that the organisation faces, grouped by relevant categories for clarity.

Description

- This provides a summary of the nature of each risk, including how it might manifest and which assets or operations may be affected.

Business impact

- This explains how each identified risk could affect the organisation's operations, revenue, regulatory compliance, and reputation, helping to prioritise response efforts.

Likelihood & severity

- This assesses the probability of occurrence and potential impact of each risk, providing a basis for risk prioritisation and mitigation planning.

This structured approach ensures that stakeholders gain clear grasp of both the current cybersecurity landscape and the critical steps required for improvement. The section facilitates informed decision-making and targeted action aligned with best practices and business objectives.

SECURITY INVESTMENTS & RETURN ON INVESTMENT

In addition to the risks identified in the previous section, here you collate data from past information-security, privacy, or cybersecurity events (including actual incidents and near-miss events) and for each identify the financial impacts. This can include assumed costs such as the amount of time taken from the team to triage, contain, and manage the event, not just transactional costs (such as purchasing new equipment).

If the organisation has not yet suffered an incident, and/or to elaborate on the potential costs, research the impacts on similar sized organisations or industry peers who have suffered an incident and disclosed the efforts and associated costs.

This can be presented in the following manner.

Event	Frequency	Experience	Estimated cost of event	Annual loss expectancy (ALE)
Name of event	How often does it occur or would you suspect the event would occur?	Has the organisation experienced this before?	Cost estimate, per event	Frequency x cost
Unauthorised access	Every 2 months	Yes: known sharing of accounts and previous external data breach; poor results in phishing campaigns	$55,000	$330,000

Event	Remediation		Estimated remediation costs	Total remediation costs
Name of event	Action/control		Cost per control	Total remediation costs
Unauthorised access	Awareness campaign		$10,000	$165,000
	Incident-response plan and testing		$45,000	
	MFA implementation		$110,000	

ROI calculation			
Operational expenses (opex) post implementation, estimated at $50,000 per year			

Year	Remediation costs	ALE
Year 1	$215,000	$330,000
Year 2	$50,000	$330,000
Year 3	$50,000	$330,000
Total potential savings over three years = $675,000		

ACTION PLAN EXAMPLE

Short-term (0–6 months):

- Implement MFA across all critical systems.
- Conduct company-wide phishing simulation.
- Update incident-response plan.

Medium-term (6–12 months):

- Complete security compliance audit.
- Strengthen vendor risk management.
- Deploy advanced threat detection.

Long-term (12+ months):

- Mature SOC capabilities.
- Implement zero-trust architecture.
- Automate security operations.

www.ingramcontent.com/pod-product-compliance
Lightning Source LLC
Chambersburg PA
CBHW071727200326
41519CB00021BC/6604